Principles of Weed Control

Principles of Weed Control

Elisabeth Crawford

SYRAWOOD
PUBLISHING HOUSE
New York

Published by Syrawood Publishing House,
750 Third Avenue, 9th Floor,
New York, NY 10017, USA
www.syrawoodpublishinghouse.com

Principles of Weed Control
Elisabeth Crawford

International Standard Book Number: 978-1-64740-009-5 (Hardback)

Cataloging-in-Publication Data

Principles of weed control / Elisabeth Crawford.
p. cm.
Includes bibliographical references and index.
ISBN 978-1-64740-009-5
1. Weeds--Control. 2. Weeds--Biological control. 3. Herbicides.
I. Crawford, Elisabeth.

SB611 .P75 2020
632.5--dc23

Printed in China.

TABLE OF CONTENTS

PREFACE

A plant which is considered to be undesirable in a particular location is termed as a weed. Weed control attempts to stop weeds from competing for resources with desired flora and fauna such as domesticated plants and livestock. There are various methods for managing weeds such as physical methods, chemical methods and biological methods. The physical methods are hand cultivation with hoes, lethal wilting with high heat and powered cultivation with cultivators. Chemical methods primarily involve the use of herbicides. There are also biological methods which employ techniques like the protection of natural predators and use of grazing animals. This textbook elucidates the concepts and innovative models around prospective developments with respect to weed control. Most of the topics introduced herein cover new techniques and the applications in this area. Those in search of information to further their knowledge will be greatly assisted by this book.

Given below is the chapter wise description of the book:

Chapter 1- Plants are the multicellular eukaryotic organisms which belong to the kingdom Plantae. Weed is an undesirable plant in a human controlled setting such as garden, farm field, park or lawn. This is an introductory chapter which will introduce briefly all the significant aspects of plants and weeds.

Chapter 2- The botanical aspect of pest control is known as weed control. Its aim is to stop weeds from contesting with desired flora and fauna. Weed control plays a vital role in agriculture and includes various types of methods like physical, chemical, cultural and biological weed control. This chapter discusses in detail the theories and methodologies related to weed control.

Chapter 3- The chemical substances which are used to control unwanted plant or weeds are known as herbicides. The classification of herbicides is done on the basis of mechanism of action or site of action. Mechanism of action refers to the first protein, enzyme or biochemical step which is affected in a plant when the herbicide is applied. All the diverse aspects of herbicides and their mechanism of action have been carefully analyzed in this chapter.

Chapter 4- Herbicides are economical, more efficient and faster than most of the other methods of weed control. However, they can cause harm to crop plants if they are not used properly. They can also have a negative effect on the environment and biodiversity of the area where they are used. All the diverse aspects of the effects of herbicide use have been carefully analyzed in this chapter.

Chapter 5- The management of weeds using multiple control tactics is known as integrated weed management. The goal of this approach towards weed control is to allow the producers to include different methods for controlling weeds in a single growing season in order to maximize the efficiency of weed control. This chapter has been carefully written to provide an easy understanding of the varied facets of integrated weed management.

At the end, I would like to thank all those who dedicated their time and efforts for the successful completion of this book. I also wish to convey my gratitude towards my friends and family who supported me at every step.

Elisabeth Crawford

Chapter 1

Plants and Weeds: An Introduction

Plants are the multicellular eukaryotic organisms which belong to the kingdom Plantae. Weed is an undesirable plant in a human controlled setting such as garden, farm field, park or lawn. This is an introductory chapter which will introduce briefly all the significant aspects of plants and weeds.

Plant

Plants are multicellular organisms in the kingdom Plantae that use photosynthesis to make their own food. There are over 300,000 species of plants; common examples of plants include grasses, trees, and shrubs. Plants have an important role in the world's ecosystems. They produce most of the world's oxygen, and are important in the food chain, as many organisms eat plants or eat organisms which eat plants. The study of plants is called botany.

Plant Characteristics

Plants are autotrophs; they produce their own food. They do so via photosynthesis, which is the process of making nutrients such as sugars from light energy and carbon dioxide. Photosynthesis occurs in cell organelles called chloroplasts, which contain chlorophyll and carotenoids, molecules that absorb light energy and change it into a usable form. Heterotrophs, on the other hand, are organisms that cannot make their own food and must eat other organisms to survive. Many heterotrophs eat plants. Other heterotrophs eat animals that have eaten plants. Plants are primary producers in many ecosystems, giving them a vital role in the survival of many other organisms. In addition, oxygen is a byproduct of photosynthesis, and many organisms depend on oxygen to survive. We couldn't live without plants.

Plants are multicellular organisms with eukaryotic cells. A eukaryotic cell is a relatively large cell with a true nucleus and other organelles that perform specific functions. Plants, protists, fungi, and animals all have eukaryotic cells. Plant cells are distinguished by their cell walls containing cellulose, chloroplasts that perform photosynthesis, and a large central vacuole that holds water and keeps the plant turgid. Prokaryotic cells, on the other hand, are small with no true nucleus or organelles except ribosomes, which produce proteins. Bacteria and archaea have prokaryotic cells.

Many plants have vascular tissue, such as xylem and phloem, that carries water and nutrients throughout the plant. This is particularly important for plants that grow upwards; water needs to travel from the roots up the stem to the leaves. Vascular tissue is found in more "complex" plants. Plants are believed to have evolved from algae-like ancestors. Today, most modern-day algae are classified as bacteria, not plants. However, green algae, which also have cellulose in their cell walls and have chloroplasts that perform photosynthesis, are sometimes grouped with plants.

Plants reproduce both sexually and asexually and have what is known as alternation of generations. A haploid stage alternates with a diploid stage. Haploid is when cells contain one set of chromosomes, while diploid is when cells contain two sets. (For reference, humans are diploid but their gametes—sperm and eggs—are haploid). In plants, two haploid gametes join to form a diploid zygote. This diploid zygote divides through mitosis to become a multicellular organism. It is called the sporophyte, and at maturity, it asexually produces haploid spores. The haploid spores then germinate into multicellular organisms called gametophytes. Gametophytes produce haploid gametes, which fuse to make a diploid organism, and the alternation between diploid and haploid starts all over again.

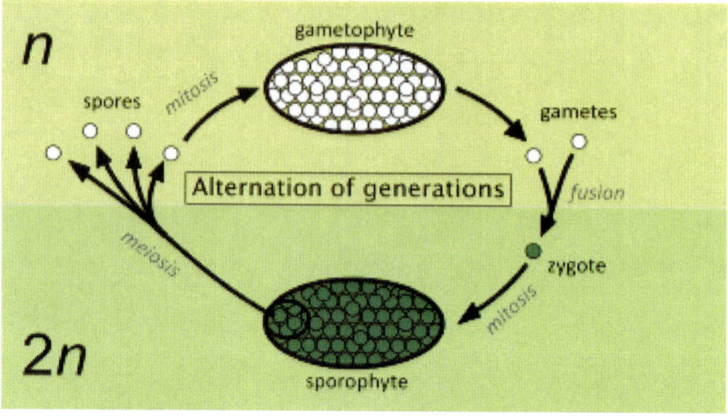

This image depicts the alternation of generations.

Plant Parts and their Functions

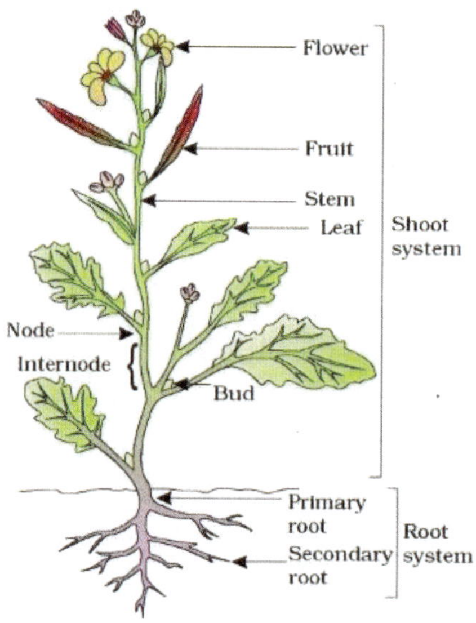

- The main functions of the root system are absorption of water and minerals from the soil, providing a proper anchorage to the plant parts, storing reserve food material and synthesis of plant growth regulators.
- In majority of the dicotyledonous plants, the direct elongation of the radicle leads to the formation of primary root which grows inside the soil. It bears lateral roots of several orders that are referred to as secondary, tertiary, etc. roots. The primary roots and its branches constitute the tap root system, as seen in the mustard plant.

- In monocotyledonous plants, the primary root is short lived and is replaced by a large number of roots. These roots originate from the base of the stem and constitute the fibrous root system, as seen in the wheat plant.
- In some plants, like grass, Monstera and the banyan tree, roots arise from parts of the plant other than the radicle and are called adventitious roots.

The Root

- The root is covered at the apex by a thimble-like structure called the root cap. It protects the tender apex of the root as it makes its way through the soil.
- Tap roots of carrot, turnip and adventitious roots of sweet potato, get swollen and store food.
- Hanging structures that support a banyan tree are called prop roots. Similarly, the stems of maize and sugarcane have supporting roots coming out of the lower nodes of the stem. These are called stilt roots.
- In some plants such as Rhizophora growing in swampy areas, many roots come out of the ground and grow vertically upwards. Such roots, called pneumatophores, help to get oxygen for respiration.

The Stem

- The region of the stem where leaves are born are called nodes while internodes are the portions between two nodes.
- Some stems perform the function of storage of food, support, protection and of vegetative propagation.
- Underground stems of potato, ginger, turmeric, zaminkand, colocasia are modified to store food in them.
- Stem tendrils which develop from axillary buds, are slender and spirally coiled and help plants to climb such as in gourds (cucumber, pumpkins, watermelon) and grapevines.
- Axillary buds of stems may also get modified into woody, straight and pointed thorns.

Thorns are found in many plants such as Citrus, Bougainvillea. They protect plants from browsing animals.

- Some plants of arid regions modify their stems into flattened (Opuntia), or fleshy cylindrical (Euphorbia) structures. They contain chlorophyll and carry out photosynthesis.
- Underground stems of some plants such as grass and strawberry, etc., spread to new niches and when older parts die new plants are formed.

The Leaf

- Leaves originate from shoot apical meristems. Leaf develops at the node and bears a bud in its axil. The axillary bud later develops into a branch.
- A typical leaf consists of three main parts: leaf base, petiole and lamina.
- The petiole help hold the blade to light. Long thin flexible petioles allow leaf blades to flutter in wind, thereby cooling the leaf and bringing fresh air to leaf surface.
- The lamina or the leaf blade is the green expanded part of the leaf with veins and veinlets. There is, usually, a middle prominent vein, which is known as the midrib. Veins provide rigidity to the leaf blade and act as channels of transport for water, minerals and food materials.

Leaf Venation

Structure of a leaf: (a) Parts of a leaf, (b) Reticulate venation, (c) Parallel venation.

- The arrangement of veins and the veinlets in the lamina of leaf is termed as venation.
- When the veinlets form a network, the venation is termed as reticulate. When the veins run parallel to each other within a lamina, the venation is termed as parallel.
- Leaves of dicotyledonous plants generally possess reticulate venation, while parallel venation is the characteristic of most monocotyledons.

Modifications of Leaves

- Leaves are often modified to perform functions other than photosynthesis. They are converted into tendrils for climbing as in peas or into spines [thorns] for defense as in cacti. The fleshy leaves of onion and garlic store food.
- Leaves of certain insectivorous plants such as pitcher plant, venus-fly trap are also modified leaves.

Root Modification	Stem Modification	Leaf Modification
Carrot and Sweet Potato.	Potato, Ginger, Turmeric, Zaminkand and Colocasia.	Onion and Garlic.

Transpiration

- Plants absorb mineral nutrients and water from the soil. Not all the water absorbed is utilised by the plant. The water evaporates through the stomata present on the surface of the leaves by the process of transpiration.
- The evaporation of water from leaves generates a suction pull (the same that you produce when you suck water through a straw) which can pull water to great heights in the tall trees. Transpiration also cools the plant.

Water Absorption through Roots can be Increased by Keeping the Plants:

- In the shade
- In dim light
- Under the fan
- Covered with a polythene bag
    - When we place a plant under the fan the speed of air flow is very high. Transpiration will take place in presence of high air flow through the stomata. Rate of transpiration increases during windy condition.
    - Increase in the rate of transpiration increases the water absorption also because when transpiration occurs, it will create a transpiration pull and more water absorption will take place.

Respiration in Plants

- In plants each part can independently take in oxygen from the air and give out carbon dioxide. Even roots can respire. Can you guess what would happen if a potted plant is overwatered?
- Plants carry out photosynthesis only during the day and respiration both during the day time as well as night.
- Did you know? For us oxygen is essential, but for those organisms which do not use it,

oxygen is toxic. In fact, our white blood cells use oxygen to kill invading bacteria. Even for humans, it may be dangerous to breathe pure oxygen for long.

The Flower

- The flower is the reproductive unit in the angiosperms. It is meant for sexual reproduction. Androecium and gynoecium are reproductive organs.
- When a flower has both androecium and gynoecium, it is bisexual. A flower having either only stamens or only carpels is unisexual.
- Aestivation: The mode of arrangement of sepals or petals in floral bud with respect to the other members of the same whorl is known as aestivation.

Androecium

- Androecium is composed of stamens. Each stamen which represents the male reproductive organ consists of a stalk or a filament and an anther. Each anther is usually bilobed and each lobe has two chambers, the pollen-sacs. The pollen grains are produced in pollen-sacs. A sterile stamen is called staminode.

Gynoecium

- Gynoecium is the female reproductive part of the flower and is made up of one or more carpels. A carpel consists of three parts namely stigma, style and ovary.
- After fertilization, the ovules develop into seeds and the ovary matures into a fruit.
- Placentation: The arrangement of ovules within the ovary is known as placentation.

The Fruit

- The fruit is a characteristic feature of the flowering plants.
- It is a mature or ripened ovary, developed after fertilisation.
- If a fruit is formed without fertilisation of the ovary, it is called a parthenocarpic fruit.
- The Seed: The ovules after fertilisation, develop into seeds.

Transport of Water and Minerals in Plants

- Plants absorb water and minerals by the roots. The roots have root hair. The root hair increase the surface area of the root for the absorption of water and mineral nutrients dissolved in water. The root hair is in contact with the water present between the soil particles.
- Can you guess how water moves from the root to the leaves? What kind of transport system is present in plants?
- Plants have pipe-like vessels to transport water and nutrients from the soil. The vessels are made of special cells, forming the vascular tissue.
- The vascular tissue for the transport of water and nutrients in the plant is called the xylem.

The xylem forms a continuous network of channels that connects roots to the leaves through the stem and branches and thus transports water to the entire plant leaves synthesise food.

- The food has to be transported to all parts of the plant. This is done by the vascular tissue called the phloem. Thus, xylem and phloem transport substances in plants.

Types of Plants

There is an incredible number of different plants in the world. Humans separate plants according to particular traits. Some of the most important differences between plants are whether they have seeds or vascular tissue. Plants have been grouped into twelve different phyla depending on these characteristics. Incidentally, learning about the types of plants also takes us on an evolutionary journey as plants emerged from aquatic systems and increased in complexity.

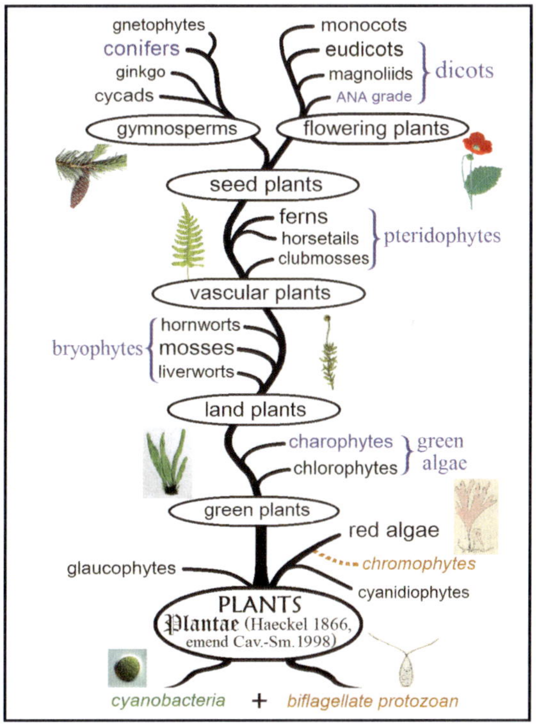

The different types of plants represented in an evolutionary tree.

Plants without Seeds

Algae: There are three different types of algae: red, green, and brown. They live in water and, for this reason, are considered primitive plants. All plants started off growing in water, and as single celled organisms. More evolutionarily advanced plants left the water. Algae are photosynthetic organisms that range from unicellular organisms to large multicellular forms.

Liverworts and hornworts: Liverworts are small plants that grow in damp environments. They do not contain the vascular tissue that transports water from the roots to leaves, which is why liverworts are usually very small and need to live in moist places. Liverworts grow simply by expanding themselves.

They do not have a true root, stem or leaves. Hornworts are similar, but have a sporophyte, which is a horn-like structure.

A liverwort.

Mosses: Mosses are close relatives of liverworts and thrive in similar environments, damp areas near water sources. However, they do not require soil to grow, which is why you can see rocks and trees covered in moss. Mosses grow apically — in other words, stems grow from their tips or other special points on the stem. Flowering plants also grow this way.

Mosses growing on rock.

Vascular Plants without Seeds

Ferns: Ferns need wet environments to reproduce so the sperm cell can swim to join with the egg cell. A new fern develops from the resulting zygote. However, ferns can survive periods with less water better than mosses and worts. Millions of years ago, ferns dominated the land and were the most common plants. There were massive fern forests.

The underside of a fern, showing the spores.

Other Types

There are several other plants that have vascular tissue but lack seeds. Whisk ferns are primitive ferns that lack typical plant organs, such as leaves. Clubmosses have branching stems with simple leaves. Horsetails, also called snake grass or puzzlegrass, are found in the only extant (still living) genus in its family. More species within the same family existed millions of years ago, including very tall trees. Club mosses and horsetails are considered fern allies since all of these plants reproduce by spores and not by seeds.

Horsetails are a very ancient type of plant.

Plants with Seeds

Gymnosperms

Cycads: Cycads are trees that like moisture and heat, therefore, they mostly grow in Central America, Africa, Southeast Asia, and Australia. They generally have long, thin leaves, and produce a cone-like structure which makes them look like palms with cones. Cycads possess a crown of large compound leaves and a thick trunk. There are only a few species left, but they were very common during the Jurassic period, which is often called the "Age of Cycads."

A cycad.

Ginkgo**:** There is currently only one ginkgo species in existence, Ginkgo biloba. It has not changed very much since the Permian period — when it covered large parts of the world — and is therefore called a living fossil. They have fan-shaped leaves, and trees are either male or female. They produce fleshy seeds that have a strong odour. Ginkgo is only found naturally in central China, but has been purposefully planted in gardens and parks around the world. Flowering plants, however, are outcompeting it.

A gingko tree.

Gnetophyta: The phylum gnetophyta is also a gymnosperm and consists of three genera that are not closely related. There are about 70 species in total. Ephedra is the largest genus, and its plants grow in deserts. Welwitschia plants grow in the desert in southwestern Africa; they have long, thin leaves. The last genus in the phylum is the namesake called Gnetum.

A plant from the Welwitschia genus.

Conifers: Conifers have woody trunks and produce cones with seeds. They grow mostly in cold northern climates and keep their leaves throughout the year. Conifers have naked seeds that are protected by cones, and the male and female cones are produced on the same tree. The pollen cones are male and produce the pollen that is spread to the female gametophyte found inside the seed cone. Seed cones are female and contain eggs on scales that form seeds when fertilized.

Conifers have seeds in protective cones.

Angiosperms

Flowering plants: Flowering plants, also called angiosperms, have male and female parts. The male parts produce pollen that is dispersed, and upon reaching the female parts produces an embryo that develops into a seed. Wind and pollinators, like bees, can pollinate these plants. Angiosperms produce flowers and fruit, and the seeds are produced and protected within this fruit. Angiosperms are divided into two groups. Monocotyledons (monocots) have one seed leaf, while dicotyledons (dicots) have two seed leaves. Monocots have parallel veins, scattered vascular tissue, and flower parts that grow in multiples of three. Dicots have net-like veins, vascular tissue in rings in the stems, flower parts that grow in multiples of 4 or 5, and are often woody. Angiosperms form the plant group most equipped to handle dry conditions, which is why they are now the most widespread plant type.

Flowering plants.

We are now living in the era of flowering plants. Evolutionarily, they are the most advanced and they make up the largest proportion of plants in the world. There are still many other diverse plant species that grow alongside them.

Plant Structure and Function

The digestibility of plants (i.e. how easy they are to digest) affects the efficiency of animals' digestive processes, and is related to the structure of plant cells and tissues.

Digestibility of Plants

Animals can easily digest the contents of plant cells, but not their cell walls. All plant cells have an outer cell wall composed of cellulose or lignin.. This reduces the amount of energy animals can extract from plants, because most animals lack the specific enzymes that would allow them to digest cellulose.

Fibre is the amount of cellulose and lignin present in the plant. In general, older and woodier plants have higher levels of cellulose and lignin. Lignin cannot be digested no matter how long it remains in the digestive system. In addition, lignin interferes with gut microbes that do have the necessary enzymes to digest cellulose, because it both acts as a physical barrier to digestion, and contains chemical bonds that cannot be broken down by normal stomach microbial flora.

'Digestibility" describes how readily the plant can be digested and its energy released for use by an animal. The digestible part of a plant includes the cell contents and the small amount of fibre that can be broken down. The amount of energy available to an animal eating a plant is determined by the proportion of fibre to cell contents in that plant.

Digestibility of Plant Structures

Grasses normally have higher fibre content than legumes such as clover. The amount of fibre in leaves of grass is twice that of legume leaves, and grass leaves are harder to digest than those of legumes. The leaves of grasses typically have a midrib made of lignin to provide support for the leaf, which adds to the higher fibre levels and lower digestibility of the leaves. The stems of most plant species have greater fibre levels compared to the leaves, and grass stems usually contain more fibre than legumes. Digestibility not only declines down the stem, but also declines more rapidly than leaf digestibility with increasing plant age.

Digestive Efficiency of Ruminants

Ruminants are better than nonruminants at digesting high-fibre diets. Leaves have a shorter rumen retention time than stems due to both faster rates of fibre digestion and higher rates of passage of undigested material. Ruminants can digest 40-50% of legume fibre and 60-70% of grass fibre. Small particles of food are digested faster than large particles because they have more surface area exposed relative to the total food volume. This is why ruminants regurgitate and re-chew (ruminate) their food after eating, to reduce the size of the food particles. Because high fibre plants are harder to break down, ruminants spend more time regurgitating and chewing grasses than legumes and more time chewing mature than immature plants.

Digestive Efficiency of Hindgut Fermenters

Hindgut fermenters such as horses do not re-chew their food. They rely on eating more food to extract enough nutrients. This means that at certain times of the year, nonruminants must eat lower quality feed in order to sustain their energy requirements. So when food becomes limited, animals such as horses and pandas must eat whatever is available, while cattle may pick and choose. In both ruminants and nonruminants efficiency of digestion drops off as dietary fibre reaches certain levels. Above 60% fibre, the efficiency of digestion in both cattle and horses drops to the point whereby not enough energy can be extracted from the food to make it worth eating. This is about the same level of fibre as the wood of many trees.

Environmental Effects of Low Digestibility

Due to the comparatively low digestibility of grass and other vegetative matter, ruminants and hindgut fermenters must consume proportionally more food than carnivores or omnivores. The high food intake and low nutrient extraction of ruminants and hindgut fermenters means that they concentrate and excrete large quantities of plant nutrients such as phosphorus and nitrogen. This leads to problems such as pollution and eutrophication of lakes and rivers when large numbers of cattle are concentrated on a comparatively small amount of farmland. Phosphorus and nitrogen from cattle effluent is washed in to lakes and waterways and causes blooms of algae that reduces water quality.

Cellulose and Lignin

The plant cell wall surrounds the cell membrane. It is made up of multiple layers of cellulose which are arranged into primary and secondary walls. Cellulose is the most common organic compound on Earth. About 33% of all plant matter is cellulose - the cellulose content of cotton is 90% and wood is 50% cellulose.

The cell walls of all vascular plants also contain a polymer called lignin. Lignin is water-resistant. It reinforces cell walls, keeping them from collapsing. This is particularly important in the xylem, because the column of water in the hollow xylem cells is under tension (negative pressure) and without the lignin reinforcement the cells would collapse.

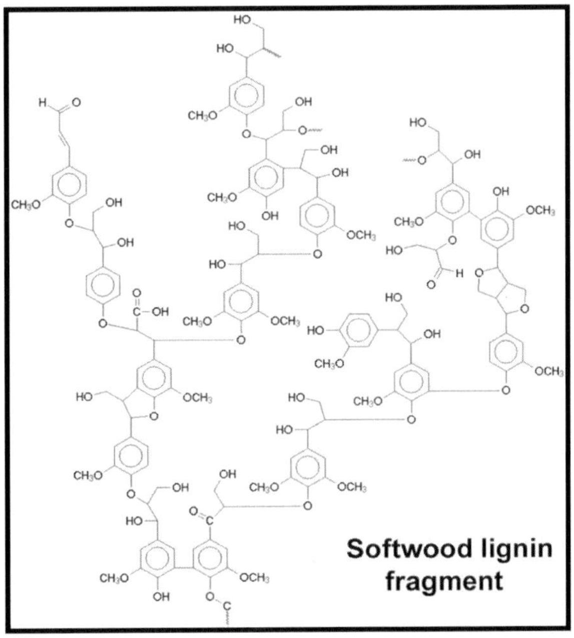

Softwood lignin fragment

Cellulose

Cellulose is a polymer made of repeating glucose molecules attached end to end (Thus cellulose is an example of a polysaccharide.). A cellulose molecule may be from several hundred to over 10,000 glucose units long. Cellulose from wood pulp has typical chain lengths between 300 and 1700 units; cotton and other plant fibres have chain lengths ranging from 800 to 10,000 units.

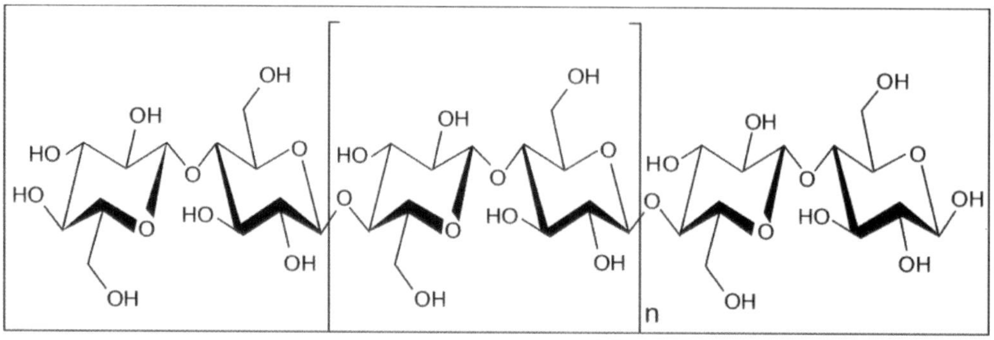

Cellulose

Cellulose is similar in form to complex carbohydrates like starch and glycogen. These polysaccharides are also made from multiple subunits of glucose. The difference between cellulose and other complex carbohydrate molecules is how the glucose molecules are linked together. In addition, cellulose is a straight chain polymer, and each cellulose molecule is long and rod-like. This differs from starch, which is a coiled molecule. A result of these differences in structure is that, compared to starch and other carbohydrates, cellulose can not be broken down into its glucose subunits by any enzymes produced by animals.

Lignin

Lignin provides the mechanical support for stems and leaves and supplies the strength and rigidity of plant walls. Lignin provides the structural strength needed by large trees to reach heights in excess of 100 m. Without lignin these trees would collapse on themselves. Also, lignin along with other cell wall constituents provides resistance to diseases, insects, cold temperatures, and other stresses. Lignin plays a crucial part in conducting water in plant stems. The structure of lignin has not been properly determined as it usually fragments upon extraction and there appears to be no consistent structure to it. The polysaccharide components of plant cell walls are highly hydrophilic and thus permeable to water, whereas lignin is more hydrophobic. The crosslinking of polysaccharides by lignin is an obstacle for water absorption to the cell wall. Thus, lignin makes it possible for the plant's vascular tissue to conduct water efficiently. Lignin is present in all vascular plants, but not in bryophytes, supporting the idea that the original function of lignin was restricted to water transport.

Plant Cells and Tissues

Cells are the smallest functional units of life. A living organism may comprise a single cell e.g. an alga; or it may be a multicellular organism made up of of billions of cells e.g a kauri tree. An individual plant contains many different cell types, each adapted to perform a particular function. However, each living plant cell is made up of the same basic components: a cell wall, plasma membrane, nucleus, and mitochondria and other organelles.

Plant Cell Structure

Each plant cell is self-contained and at least partially self-sufficient. This is because plant cells have a number of organelles that perform functions such as storing food, processing waste, fixing and releasing chemical energy, and copying their genetic material. The nucleus, the mitochondrion, the plastids, the Golgi apparatus, and the endoplasmic reticulum are all examples of organelles. Some organelles, such as mitochondria and chloroplasts, have their own genetic material separate from that found in the nucleus of the cell. Such organelles are thought to have originated through the process of endosymbiosis. Other structures such as the cell wall and the cell membrane serve to protect the cell and maintain the shape, structure, and functioning of the cell.

The Plant Cell Wall

Unlike animal cells plant cells have an outer cell wall composed of polysaccharides, in particular cellulose. Cell walls provide rigidity and mechanical support, maintain cell shape and the direction of cell growth. The cell wall also prevents expansion when water enters the cell.

Cell Wall Structure

The cell wall of a plant cell is made up of primary and secondary cell walls. When one cell divides into two, a primary cell wall forms around each two new cells. The primary cell wall is laid down just outside the cell membrane, and is flexible, allowing the new cells to grow. This new cell wall is mostly made of cellulose molecules, arranged into thin hair-like strands called microfibrils. The microfibrils are arranged in a meshwork pattern along with other components such as hemicellulose, glycans and pectins, which link them together and help strengthen the cell wall.

The secondary cell wall is constructed between the plasma membrane and the primary cell wall after the cell has finished growing. It is made by laying down successive layers of cellulose microfibrils and lignins. Mature xylem cells are heavily lignified - they make up the 'wood' of woody plants.

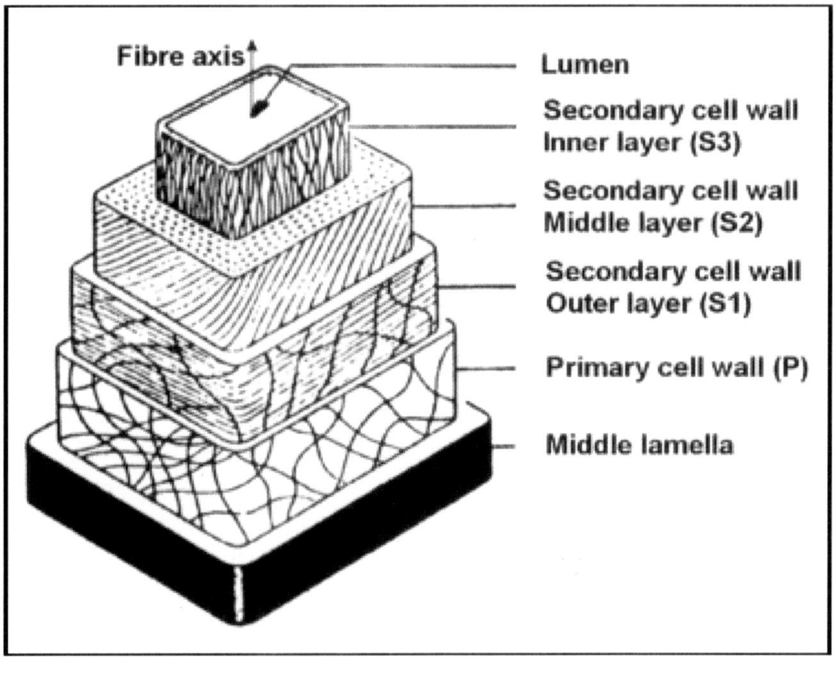

Cell Plasma Membrane

Every living cell is bounded by an outer membrane called the cell plasma membrane, or plasmalemma. The cell membrane separates the cell's contents from the surrounding environment and controls the movement of materials into and out of the cell.

The cell membrane is made up of two phospholipid layers (a bi-layer)which spontaneously arrange so that the hydrophobic "tail" regions are protected from the surrounding water, causing the more hydrophilic "head" regions to be pointed towards either the cytoplasm or the exterior of the cell. Substances such as amino acids, nucleic acids, carbohydrates, proteins, and many ions are unable to diffuse across the membrane but must enter or leave the cell through active transport, under the control of the cell membrane.

This involves some of the many different proteins and lipids embedded in the plasma membrane. Other proteins are involved in other cellular processes such as cell linkage, ion channel flow and cell communication. The plasma membrane also serves as the attachment point for both the intracellular cytoskeleton (found in all eukaryote cells) and the cell wall.

Cytoplasm

The cytoplasm is the gooey, semi-transparent fluid in which the other organelles are suspended. Processes such as breaking down food molecules into smaller molecules happens in the cytoplasm.

The cytoplasm consists mostly of water, dissolved ions, small molecules, and large water-soluble molecules such as proteins. It also has a high concentration of potassium ions and a low concentration of sodium ions, which help in controlling the amount of water in the cell. If you make a wet mount of a leaf from the pondweed Elodea, you may be able to see the cell's organelles, especially the chloroplasts, moving as the cytoplasm flows through the cell.

Nucleus

In eukaryote cells the nucleus is surrounded by a double membrane called the nuclear envelope. This separates the genetic material from the rest of the cell. The nuclear envelope has a large number of holes or 'pores' that allow the cell to move molecules across the nuclear envelope and in and out of the nucleus. These molecules include RNA and ribosomes moving from nucleus to the cytoplasm, and proteins, carbohydrates, and lipids moving into the nucleus. You could say that these molecules provide a means of communication between the cytoplasm and the nucleus. The nucleus performs two important functions: it controls the activity of the cell by determining what proteins are produced and when they are produced by the cell; and it also stores the genetic information of the cell which is then passed on to daughter cells during cell division.

Vacuole

The vacuoles of most plant cells are large, filling much of the space inside the cell wall and pushing the cytoplasm out to the periphery of the cell.

A single, large vacuole is found in the cytoplasm of most mature plant cells. (Animal cells have much smaller, multiple vacuoles.) The vacuole is bounded by a membrane that keeps the vacuole's contents separate from the cytoplasm. Young plant cells typically contain a number of small independent vacuoles. These merge together to form one large vacuole as the cell matures, so that in a fully mature cell the vacuole may occupy more than 90% of the cell volume.

Vacuoles typically contain salts, sugars, and sometimes proteins,a nd have a variety of secretory, excretory, and storage functions. They are usually slightly acidic. Some of them, like those in grapefruit and lemons are very acidic – which is where the sour taste of these fruit comes from. The vacuole may also be involved in the breakdown of old organelles or macromolecules and then recycling the products back to the rest of the cell. For example, an entire mitochondrion may be engulfed by the vacuole, so that it is surrounded by a 'bubble' of the vacuole membrane. This 'bubble' is then pinched off, forming a vesicle suspended in the vacuole. After the mitochondrion has been digested the vesicle disappears.

Endoplasmic Reticulum

The endoplasmic reticulum (often abbreviated to ER) is an extensive network of membranous sack-like folds, tubes and vesicles, held together by the cytoskeleton. This network is continuous with the nuclear membrane. Plant cells contain both rough and smooth ER. The relative proportions of rough and smooth endoplasmic reticulum in a cell can change quickly, depending on the cell's metabolic needs. The endoplasmic reticulum provides a site for the synthesis of lipids and proteins and for the storage and transport of those molecules.

Rough Endoplasmic Reticulum

The surface of the rough endoplasmic reticulum is studded with ribosomes, giving it a "rough" appearance. This means that the rough ER is a site of protein synthesis. Once produced, the proteins are packaged into vesicles and transported to the Golgi apparatus, where they are modified and packaged for export from the cell or transport to where they will be used.

Smooth Endoplasmic Reticulum

Smooth endoplasmic reticulum is also made up of tubules and vesicles that branch forming a network attached to the nuclear membrane. The smooth endoplasmic reticulum provides an increased surface area for the action or storage of key enzymes and the products of these enzymes. The smooth endoplasmic reticulum also produces lipids, such as the phospholipids that make up the cell membrane - in fact, it is the source of new cell membrane material. It also metabolises carbohydrates breaking them down into glucose providing energy for the cell.

The smooth ER is also involved in the breakdown and detoxification of drugs and chemicals e.g. in human liver cells smooth ER is involved in the metabolism of ethanol. Long-term use of alcohol leads to a increased amount of smooth ER in the cells, which increases the rate of alcohol metabolism. This explains why people who regularly drink large amounts of alcohol often have a high tolerance for alcohol.

Golgi Apparatus

The Golgi apparatus is a collective term for Golgi bodies: flattened, membrane-enclosed compartments arranged into a stack of five or six pancake-like structures, often with globs or cisternae at the ends. The Golgi bodies are processing and sorting centres for cellular materials such as proteins, enzymes and lipids. Cellular materials are transported from production areas such as the endoplasmic reticulum to the Golgi bodies and, after sorting, to cisternae at the ends of the Golgi bodies. At the cisternae the molecules are enclosed in a phospholipid membrane. This membrane is then pinched off to form a vesicle, which transports the molecules to their destination and fuses with a membrane there to release its contents. In plants, cellulose sub-units are packaged into vesicles by the Golgi bodies, which then travel to and fuse with the plasma membrane. The vesicles discharge their contents to the outside of the cell and the cellulose is incorporated in the cell wall.

Mitochondria

Mitochondria - the cell's 'power plants' - generate most of the cell's supply of the energy-carrying molecule adenosine triphosphate (ATP). MItochondria are also involved in a range of other processes: signalling, cellular differentiation, cell death, control of the cell cycle, and cell growth. They evolved as the result of endosymbiosis from ancient bacteria, which were engulfed by the ancestors of eukaryote cells more than a billion years ago.

Mitochondria are much smaller than plastids - around 0.5μm in diameter i.e. too small to observe with a light microscope. The number of mitochondria in a cell varies widely: from one to several thousand, depending on organism and tissue type. Mitochondria replicate (by binary fission) mainly in response to the energy needs of the cell. In other words, their growth and division is not linked to the cell cycle. When the energy needs of a cell are high, mitochondria grow and divide. When energy use is low, mitochondria are destroyed or become inactive.

Mitochondria Structure

A mitochondrion is composed of inner and outer membranes separated by a region called the intermembrane space. The inner membrane is highly folded, forming projections called cristae. The membrane folds greatly increase the surface area of the inner membrane, which is studded with the enzymes involved in making ATP. The area inside the inner membrane is called the matrix. The matrix contains enzymes that produce molecules used in the production of ATP on the inner membrane. The mitochondrial matrix also contains the mitochondrion's DNA and ribosomes.

Mitochondria and Aerobic Respiration

The primary role of the mitochondrion is the production of ATP, used to power cellular processes. Glycolysis in the cytoplasm produces pyruvate and NADH. Once in the mitochonrion, pyruvate is oxidised in the Krebs cycle to produce CO_2, a small amount of ATP, and hydrogen ions (H^+). These then enter the Electron Transport Chain, which generates a large amount of ATP, and water (through the combination of the H^+ ions with oxygen). In other words, if oxygen is not available other less efficient anaerobic processes, such as fermentation, must be used to generate ATP.

Aerobic respiration has a much higher yield of ATP than anaerobic respiration. For example, one molecule of glucose will produce between 32 to 36 molecules of ATP under aerobic conditions; compared to 2 molecules of ATP under anaerobic conditions.

Plastids

Like vacuoles and cell walls, plastids are characteristic of plant cells. Plastids are responsible for photosynthesis, storage of products like starch and for the synthesis of many classes of molecules, such as fatty acids and terpenes, which are needed as cellular building blocks and for normal functions of the plant. Like mitochondria, plastids originated through endosymbiosis. In plants, plastids may differentiate into several forms, depending upon which function they need to play in the cell.

Chloroplasts

Chloroplasts are give plants their characteristic green colour and are the site of photosynthesis. Like a mitochondrion, a chloroplast has an outer and inner membrane, with an intermembrane space between them. Within the inner membrane lies the stroma, which contains the thylakoid membranes - the site of photosynthesis. The thylakoid membranes form a series of flattened, fluid-filled, interconnected tubules that are often stacked on top of each other to form a structure called a granum. The photosynthetic pigments, such as chlorophyll, are located in the thylakoid membranes.

Chromoplasts

Chromoplasts - 'coloured bodies' - synthesise and store pigments. They are found in coloured organs of plants such as fruit and flower petals, giving them their distinctive colours. When fruit such as tomatoes ripen, the green chloroplasts to are converted to chromoplasts that mostly contain red carotenoids.

Amyloplasts

During the day, when plants are photosynthesising, amyloplasts store glucose and convert it into a form of starch for storage. They can also convert the starch back into sugar, when the plant needs energy.

Statoliths

Statoliths are a specialised form of amyloplast used by plants in detecting and responding to gravity. Statoliths are denser than the cytoplasm and tend to move towards the bottom of the cell, indicating which direction is 'down'. They are found mainly in root tissues.

Plant Tissues

In plants, the processes of differentiation and specialisation allow groups of cells that are structurally or functionally distinct to form tissues that have specific functions. Tissues composed of only one type of cell are called simple tissues, while those made up of two or more types of cells are called complex tissues. Parenchyma and collenchyma are simple tissues; xylem, phloem, and epidermis are complex tissues. Tissues can also be grouped into tissue systems, and in plants these are ground tissue,vascular tissue, and dermal tissue. Ground tissue is made up of parenchyma, collenchyma and sclerenchyma. Vascular tissue system consists of the two conducting tissues – xylem and phloem. Dermal tissue comprises the epidermis: pavement cells, guard cells, and trichomes. (In older woody plants the epidermis is replaced by the periderm, or bark).

Ground Tissue

Most of a plant's body is made up of ground tissue, which has a variety of functions, including photosynthesis, storage of carbohydrates and mechanical support of the plant. Ground tissue can be divided into three tissue types; parenchyma, collenchyma and sclerenchyma.

Parenchyma

Parenchyma is the most common and versatile ground tissue: it forms the cortex and pith of stems, the cortex of roots, the mesophyll of leaves, the pulp of fruits, and the endosperm of seeds. Parenchyma cells are living cells, and may still be able to divide when they are mature. Because of this, parenchyma cells are important for regeneration and wound healing. They have thin but flexible cell walls because the secondary cell wall is usually absent. Photosynthesis usually occurs in parenchyma cells (the mesophyll), and they function in the storage of carbohydrates. They are generally polygonal when packed close together, but are nearly spherical when separated from their neighbours. They have large central vacuoles, which allow the cells to store and regulate ions, waste products and water.

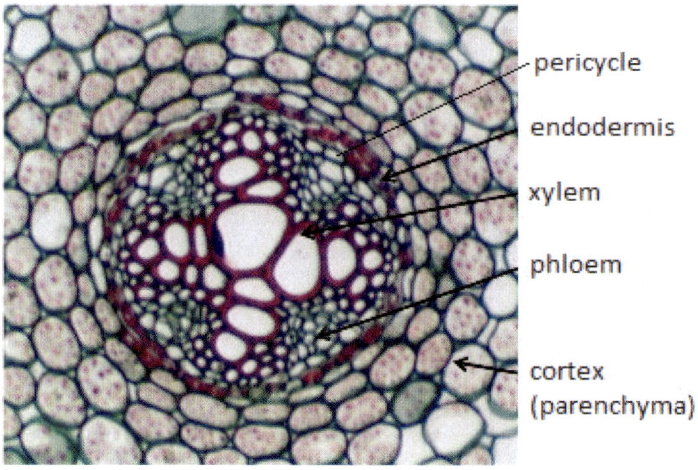

Cross-section of a dicot root, showing the vascular tissue surrounded by large, rounded, thin-walled parenchyma cells.

Collenchyma

Collenchyma cells are elongated and have unevenly thickened walls. Like parenchyma cells, they are still alive when mature. Collenchyma provides structural support, particularly in growing shoots and leaves. The stringy pieces in the stalks of celery, for example, are composed of collenchyma.

Sclerenchyma

Mature sclerenchyma is composed of dead cells with extremely thick secondary walls cell walls composed of cellulose and lignin: these walls make up to 90% of the whole cell volume. Sclerenchyma tissue is a supporting tissue. There are two types of sclerenchyma cells, fibres and sclereids. Fibres are generally long, slender cells, which commonly occur in strands or bundles, and can be seen in the diagram below as dense red caps to the outside of the vascular tissue. Fibres are of great economical importance, since they make up the source material for many fabrics such as flax, hemp and jute. Sclereids are variable in shape and usually shorter than fibre cells. They make up the seed coats of seeds, the shells of nuts and the stone of stone fruits.

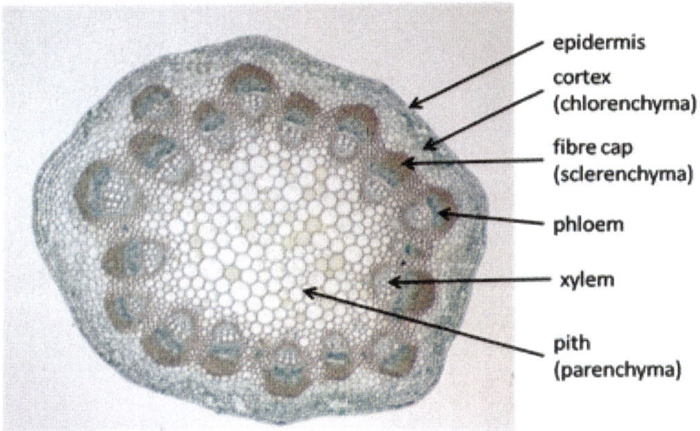

Cross-section of a dicot stem, showing cortex, pith, and vascular tissue.

Vascular Tissue

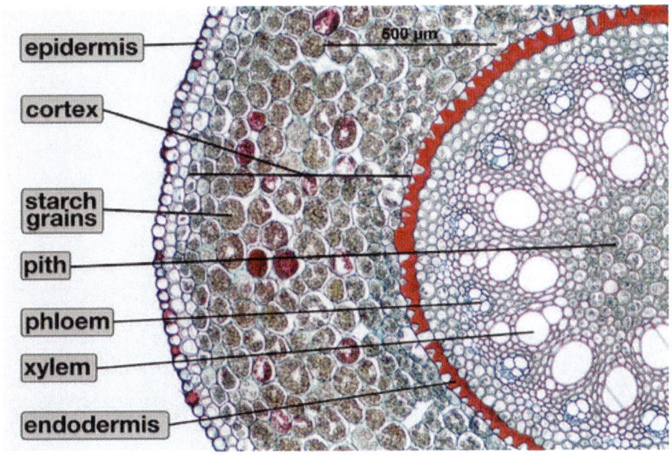

Cross-section of a monocot root, showing cortex, pith and vascular tissue.

Vascular tissue is a complex tissue and characteristic of vascular plants. The two main components of vascular tissue are the xylem and phloem. The vascular tissue in plants is arranged in long, discrete strands called vascular bundles. These bundles include xylem and phloem, as well as supporting and protective cells. The arrangement of vascular tissue differs in monocot and dicot plants, as you can see by comparing the two previous images with the two that follow below.

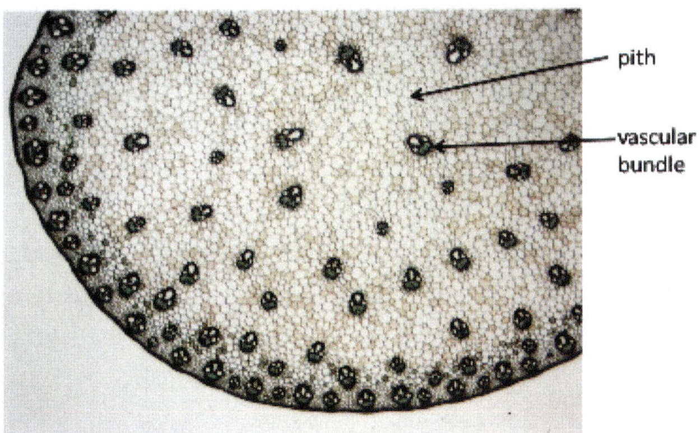

Cross-section through a monocot stem, showing pith and vascular bundles.

Xylem

The xylem forms a continuous system that transports water and dissolved mineral nutrients throughout the plant from the roots. There are two types of conducting cells in the xylem: tracheids and vessel elements. (Xylem also contains parenchyma and sclerenchyma cells.) Both tracheids and vessel elements are dead when mature and have thick, lignified cell walls.

Tracheids are long, tapering cells with pits in their walls, through which the water passes as it moves up the plant. Vessels have a larger diameter than tracheids and their end walls have disappeared. They are stacked end-on-end to form a continuous series of tubes up the stem, allowing for more efficient water movement than in tracheids.

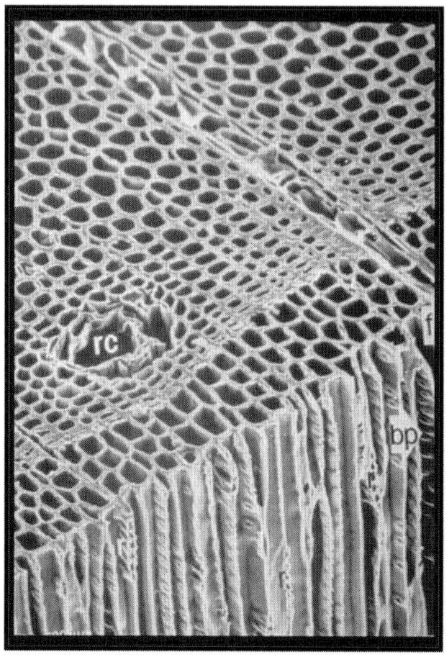

Section through woody tissue, showing xylem tracheids.

Phloem

Phloem transports sugars, proteins and minerals around the plant. Unlike the xylem where flow is only one way, from the roots up, phloem can move sugars both up and down the stem. In flowering plants, phloem is contains four cell types: sieve tube members, companion cells, fibres and parenchyma. Sieve tube members are very highly specialised. The ends of the cell wall are perforated by sieve plates, and sap passes through the sieve plates from one sieve tube member to another. Mature sieve tube members are not dead, but they do lose their nuclei and most of their cytoplasm in order to accommodate the flow of phloem sap once they mature. For this reason, mature sieve tube members have accompanying companion cells which supply the RNA and proteins that keep the sieve tube member cell alive. Companion cells are also involved in the off-loading and on-loading of sugars from the phloem sap. As with the xylem tissue the fibre cells provide structural support to the tissue and parenchyma cells provide storage for various substances.

Dermal Tissue

The dermal tissue is the boundary between the plant and the outside world. It functions in protection against water loss, regulation of gas exchange, secretion, and (especially in roots) absorption of water and mineral nutrients. There are two main types of dermal tissue: epidermis and periderm. The epidermis is usually transparent (epidermal cells lack chloroplasts) and coated on its outer surface with a waxy cuticle that prevents water loss. Epidermal tissue includes several differentiated cell types, including pavement cells, guard cells, and trichomes (epidermal hairs). The other dermal tissue type is the periderm, or bark. It replaces the epidermis of stems and roots once a plant has developed secondary thickening.

Epidermis

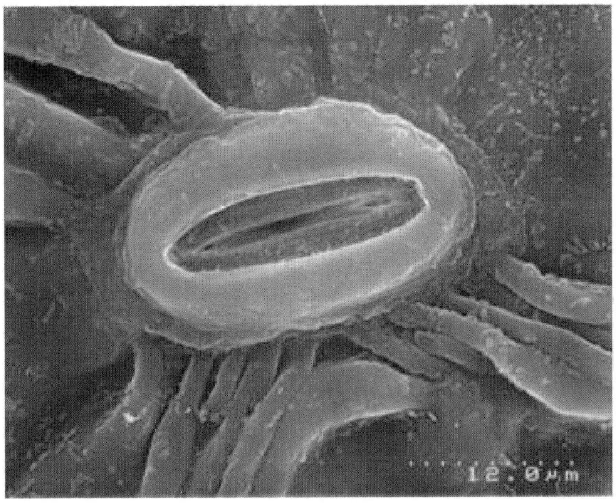

Carmichaelia stoma.

Pavement cells are the most numerous epidermal cells, irregularly shaped and the least specialised. They do not contain chlorophlasts. Because they are transparent, light can penetrate into the interior of stems and leaves, where photosynthesis occurs. Pavement cells are tightly packed together to prevent water loss. The epiderms if leaves and stems is covered by a waxy cuticle, which also helps prevent water loss. However, the epidermis is also perforated by openings called stomata, which permit gas exchange. Specialised cells called guard cells control the opening and closing of these stomata, regulating the exchange of gases and water vapour between the outside air and the interior of the leaf. A stoma is open when the guard cells surrounding it are swollen (turgid) with water, and closed when the guard cells lose water and become flaccid.

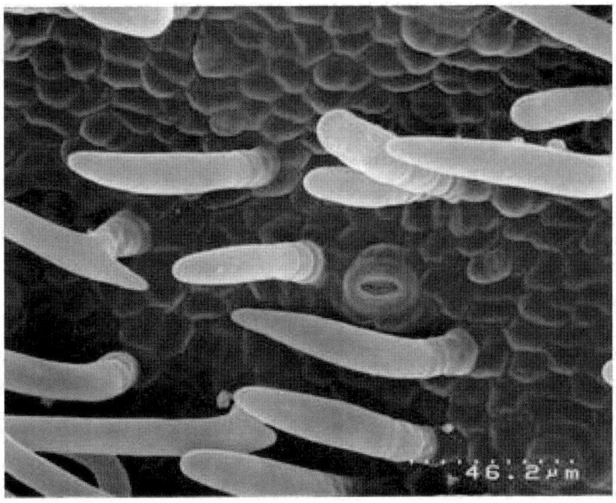

Pittosporum trichomes and stoma.

Trichomes are minute hairs on the epidermis. On leaves these hairs can interfere with the feeding of some herbivores, prevent frost forming on and damaging the leaf, reduce evaporation due to wind, and reflect light. And in locations where plants get most of their water from cloud drip, leaf hairs maximise this process by acting as points of condensation. Hairs on the roots increase the surface area available to the plant for uptake of water and minerals.

Plant Growth and Development

Growth

Growth is regarded as one of the most fundamental and conspicuous characteristics of a living being. Growth can be defined as an irreversible permanent increase in size of an organ or its parts or even of an individual cell. Generally, growth is accompanied by metabolic processes (both anabolic and catabolic), that occur at the expense of energy. Therefore, for example, expansion of a leaf is growth.

Plant Growth Generally is Indeterminate

Plant growth is unique because plants retain the capacity for unlimited growth throughout their life. This ability of the plants is due to the presence of meristems at certain locations in their body. The cells of such meristems have the capacity to divide and self-perpetuate. The product, however, soon loses the capacity to divide and such cells make up the plant body. This form of growth wherein new cells are always being added to the plant body by the activity of the meristem is called the open form of growth. Root apical meristem and shoot apical meristem are responsible for the primary growth of the plants and principally contribute to the elongation of the plants along their axis. You also know that in dicotyledonous plants and gymnosperms, the lateral meristems, vascular cambium and cork-cambium appear later in life. These are the meristems that cause the increase in the girth of the organs in which they are active. This is known as secondary growth of the plant.

Growth is Measurable

Growth, at a cellular level, is principally a consequence of increase in the amount of protoplasm. Since increase in protoplasm is difficult to measure directly, one generally measures some quantity which is more or less proportional to it. Growth is, therefore, measured by a variety of parameters some of which are: increase in fresh weight, dry weight, length, area, volume and cell number. You may find it amazing to know that one single maize root apical mersitem can give rise to more than 17,500 new cells per hour, whereas cells in a watermelon may increase in size by upto 3,50,000 times. In the former, growth is expressed as increase in cell number; the latter expresses growth as increase in size of the cell. While the growth of a pollen tube is measured in terms of its length, an increase in surface area denotes the growth in a dorsiventral leaf.

Phases of Growth

The period of growth is generally divided into three phases, namely, meristematic, elongation and maturation. Let us understand this by looking at the root tips. The constantly dividing cells, both at the root apex and the shoot apex, represent the meristematic phase of growth. The cells in this region are rich in protoplasm, possess large conspicuous nuclei. Their cell walls are primary in nature, thin and cellulosic with abundant plasmodesmatal connections. The cells proximal (just next, away from the tip) to the meristematic zone represent the phase of elongation. Increased vacuolation, cell enlargement and new cell wall deposition are the characteristics of the cells in this phase. Further away from the apex, i.e., more proximal to the phase of elongation, lies the portion of axis which is undergoing the phase of maturation. The cells of this zone, attain their maximal size in terms of wall thickening and protoplasmic modifications.

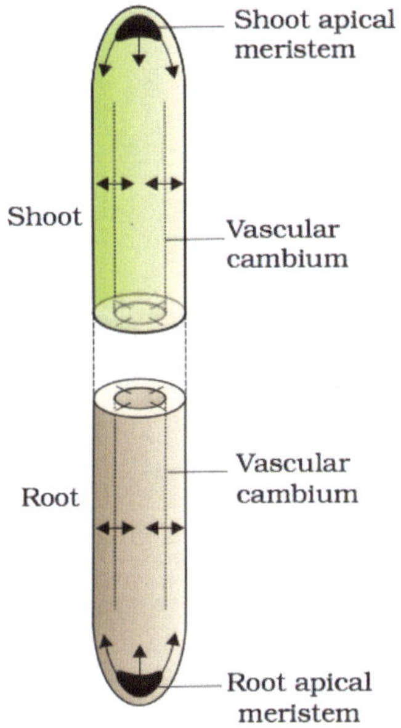

Diagrammatic representation of locations of root apical meristem, shoot aplical meristem and vascular cambium. Arrows exhibit the direction of growth of cells and organ.

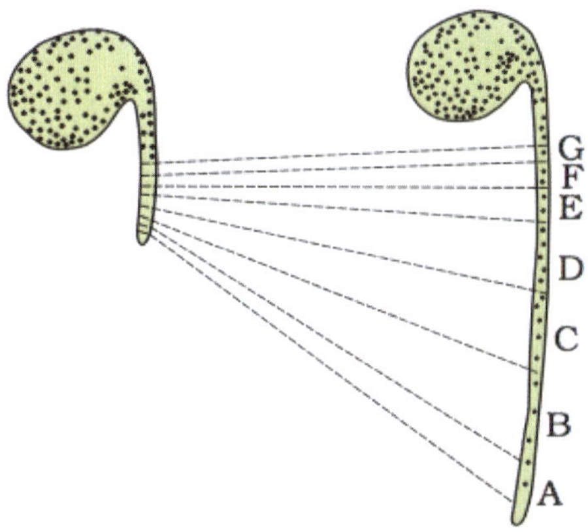

Detection of zones of elongation by the parallel line technique. Zones A, B, C, D immediately behind the apex have elongated most.

Growth Rates

The increased growth per unit time is termed as growth rate. Thus, rate of growth can be expressed mathematically. An organism, or a part of the organism can produce more cells in a variety of ways.

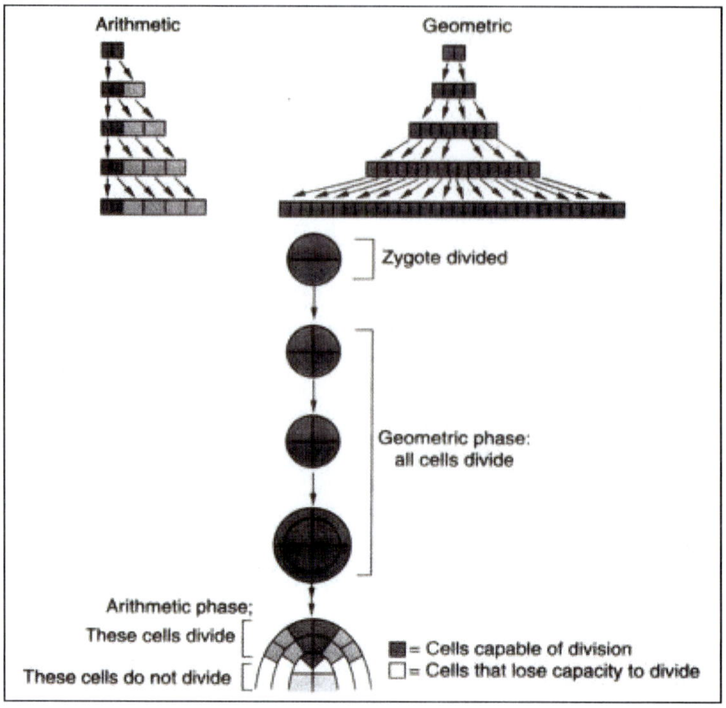

Diagrammatic representation of : (a) Arithmetic (b) Geometric growth and (c) Stages during embryo development showing geometric and arithematic phases.

The growth rate shows an increase that may be arithmetic or geometrical. In arithmetic growth, following mitotic cell division, only one daughter cell continues to divide while the other differentiates and matures. The simplest expression of arithmetic growth is exemplified by a root elongating at a constant rate. On plotting the length of the organ against time, a linear curve is obtained. Mathematically, it is expressed as:

$L_t = L_o + rt$

L_t = length at time 't'

L_o = length at time 'zero'

r = growth rate / elongation per unit time.

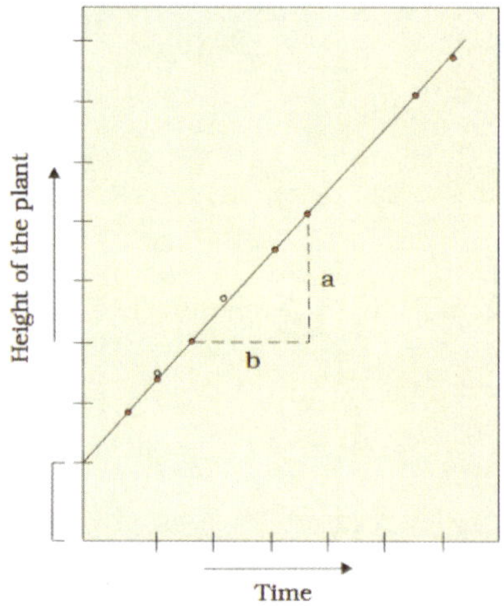

Constant linear growth, a plot of length L against time t.

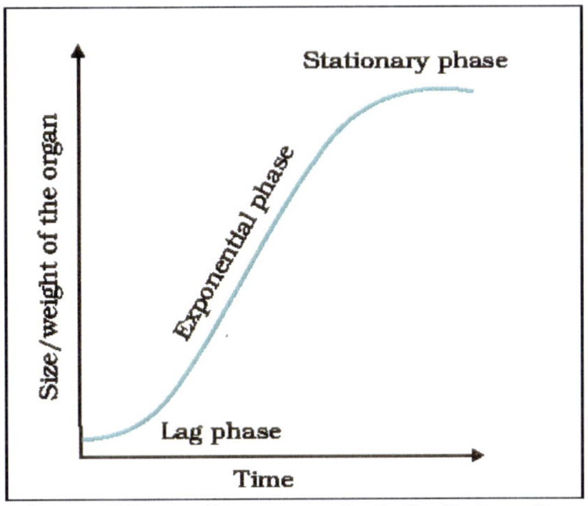

An idealised sigmoid growth curve typical of cells in culture, and many higher plants and plant organs.

Let us now see what happens in geometrical growth. In most systems, the initial growth is slow (lag phase), and it increases rapidly thereafter – at an exponential rate (log or exponential phase). Here, both the progeny cells following mitotic cell division retain the ability to divide and continue to do so. However, with limited nutrient supply, the growth slows down leading to a stationary phase. If we plot the parameter of growth against time, we get a typical sigmoid or S-curve. A sigmoid curve is a characteristic of living organism growing in a natural environment. It is typical for all cells, tissues and organs of a plant. Can you think of more similar examples? What kind of a curve can you expect in a tree showing seasonal activities?

The exponential growth can be expressed as:

$$W_1 = W_0 \ e^{rt}$$

W_1 = final size (weight, height, number etc.)

W_0 = initial size at the beginning of the period

r = growth rate

t = time of growth

e = base of natural logarithms

Here, r is the relative growth rate and is also the measure of the ability of the plant to produce new plant material, referred to as efficiency index. Hence, the final size of W_1 depends on the initial size, W_0.

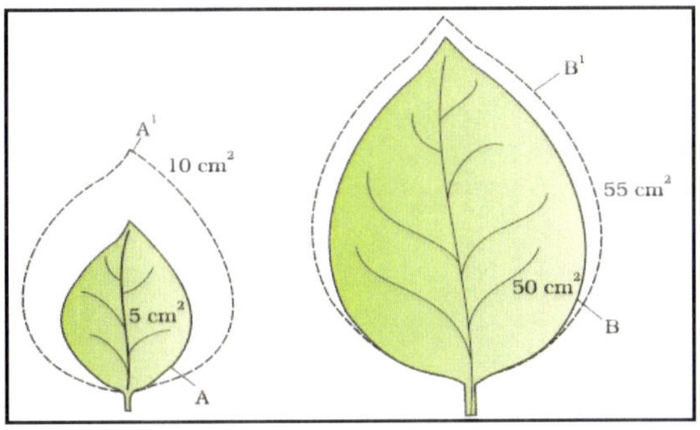

Diagrammatic comparison of absolute and relative growth rates. Both leaves A and B have increased their area by 5 cm^2 in a given time to produce A^1, B^1 leaves.

Quantitative comparisons between the growth of living system can also be made in two ways : (i) measurement and the comparison of total growth per unit time is called the absolute growth rate. (ii) The growth of the given system per unit time expressed on a common basis, e.g., per unit initial parameter is called the relative growth rate. In two leaves, A and B, are drawn that are of different sizes but shows absolute increase in area in the given time to give leaves, A1 and B1 . However, one of them shows much higher relative growth rate.

Conditions for Growth

Why do you not try to write down what you think are necessary conditions for growth? This list may have water, oxygen and nutrients as very essential elements for growth. The plant cells grow in size by cell enlargement which in turn requires water. Turgidity of cells helps in extension growth. Thus, plant growth and further development is intimately linked to the water status of the plant. Water also provides the medium for enzymatic activities needed for growth. Oxygen helps in releasing metabolic energy essential for growth activities. Nutrients (macro and micro essential elements) are required by plants for the synthesis of protoplasm and act as source of energy.

In addition, every plant organism has an optimum temperature range best suited for its growth. Any deviation from this range could be detrimental to its survival. Environmental signals such as light and gravity also affect certain phases/stages of growth.

Differentiation, Dedifferentiation and Redifferentiation

The cells derived from root apical and shoot-apical meristems and cambium differentiate and mature to perform specific functions. This act leading to maturation is termed as differentiation. During differentiation, cells undergo few to major structural changes both in their cell walls and protoplasm. For example, to form a tracheary element, the cells would lose their protoplasm. They also develop a very strong, elastic, lignocellulosic secondary cell walls, to carry water to long distances even under extreme tension. Try to correlate the various anatomical features you encounter in plants to the functions they perform.

Plants show another interesting phenomenon. The living differentiated cells, that by now have lost the capacity to divide can regain the capacity of division under certain conditions. This phenomenon is termed as dedifferentiation. For example, formation of meristems – interfascicular cambium and cork cambium from fully differentiated parenchyma cells. While doing so, such meristems/tissues are able to divide and produce cells that once again lose the capacity to divide but mature to perform specific functions, i.e., get redifferentiated. List some of the tissues in a woody dicotyledenous plant that are the products of redifferentiation. How would you describe a tumour? What would you call the parenchyma cells that are made to divide under controlled laboratory conditions during plant tissue culture?

The growth in plants is open, i.e., it can be indeterminate or determinate. Now, we may say that even differentiation in plants is open, because cells/tissues arising out of the same meristem have different structures at maturity. The final structure at maturity of a cell/tissue is also determined by the location of the cell within. For example, cells positioned away from root apical meristems differentiate as root-cap cells, while those pushed to the periphery mature as epidermis.

Development

Development is a term that includes all changes that an organism goes through during its life cycle from germination of the seed to senescence. Diagrammatic representation of the sequence of processes which constitute the development of a cell of a higher plant is given in figure. It is also applicable to tissues/organs.

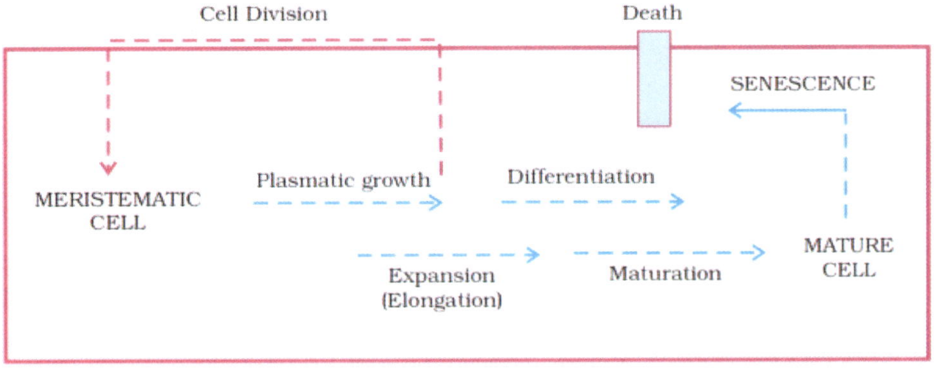

Sequence of the developmental process in a plant cell.

Plants follow different pathways in response to environment or phases of life to form different kinds of structures. This ability is called plasticity, e.g., heterophylly in cotton, coriander and larkspur. In such plants, the leaves of the juvenile plant are different in shape from those in mature plants. On the other hand, difference in shapes of leaves produced in air and those produced in water in buttercup also represent the heterophyllous development due to environment. This phenomenon of heterophylly is an example of plasticity.

Heterophylly in (a) larkspur and (b) buttercup.

Thus, growth, differentiation and development are very closely related events in the life of a plant. Broadly, development is considered as the sum of growth and differentiation. Development in plants (i.e., both growth and differentiation) is under the control of intrinsic and extrinsic factors. The former includes both intracellular (genetic) or intercellular factors (chemicals such as plant growth regulators) while the latter includes light, temperature, water, oxygen, nutrition, etc.

Plant Growth Regulators

Characteristics

The plant growth regulators (PGRs) are small, simple molecules of diverse chemical composition. They could be indole compounds (indole-3-acetic acid, IAA); adenine derivatives (N^6-furfurylamino purine, kinetin), derivatives of carotenoids (abscisic acid, ABA); terpenes (gibberellic acid, GA_3) or gases (ethylene, C_2H_4). Plant growth regulators are variously described as plant growth substances, plant hormones or phytohormones in literature.

The PGRs can be broadly divided into two groups based on their functions in a living plant body. One group of PGRs are involved in growth promoting activities, such as cell division, cell enlargement, pattern formation, tropic growth, flowering, fruiting and seed formation. These are also called plant growth promoters, e.g., auxins, gibberellins and cytokinins. The PGRs of the other group play an important role in plant responses to wounds and stresses of biotic and abiotic origin. They are also involved in various growth inhibiting activities such as dormancy and abscission. The PGR abscisic acid belongs to this group. The gaseous PGR, ethylene, could fit either of the groups, but it is largely an inhibitor of growth activities.

The Discovery of Plant Growth Regulators

Interestingly, the discovery of each of the five major groups of PGRs have been accidental. All this started with the observation of Charles Darwin and his son Francis Darwin when they observed that the coleoptiles of canary grass responded to unilateral illumination by growing towards the light source (phototropism). After a series of experiments, it was concluded that the tip of coleoptile was the site of transmittable influence that caused the bending of the entire coleoptile. Auxin was isolated by F.W. Went from tips of coleoptiles of oat seedlings. the tip of coleoptile was esite of transmittable influence that caused the bending of the entire coleoptile. Auxin was isolated by F.W. Went from tips of coleoptiles of oat seedlings.

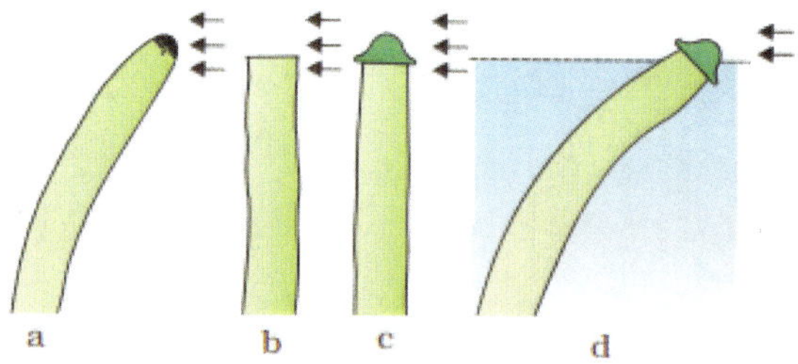

Experiment used to demonstrate that tip of the coleoptile is the source of auxin. Arrows indicate direction of light

The 'bakane' (foolish seedling) disease of rice seedlings, was caused by a fungal pathogen Gibberella fujikuroi. E. Kurosawa reported the appearance of symptoms of the disease in uninfected rice seedlings when they were treated with sterile filtrates of the fungus. The active substances were later identified as gibberellic acid.

F. Skoog and his co-workers observed that from the internodal segments of tobacco stems the callus (a mass of undifferentiated cells) proliferated only if, in addition to auxins the nutrients medium was supplemented with one of the following: extracts of vascular tissues, yeast extract, coconut milk or DNA. Skoog and Miller, later identified and crystallised the cytokinesis promoting active substance that they termed kinetin.

During mid-1960s, three independent researches reported the purification and chemical characterisation of three different kinds of inhibitors: inhibitor-B, abscission II and dormin. Later all the three were proved to be chemically identical. It was named abscisic acid (ABA).

Cousins confirmed the release of a volatile substance from ripened oranges that hastened the ripening of stored unripened bananas. Later this volatile substance was identified as ethylene, a gaseous PGR.

Physiological Effects of Plant Growth Regulators

Auxins

Auxins was first isolated from human urine. The term 'auxin' is applied to the indole-3-acetic acid (IAA), and to other natural and synthetic compounds having certain growth regulating properties. They are generally produced by the growing apices of the stems and roots, from where they migrate to the regions of their action. Auxins like IAA and indole butyric acid (IBA) have been isolated from plants. NAA (naphthalene acetic acid) and 2, 4-D (2, 4-dichlorophenoxyacetic) are synthetic auxins. All these auxins have been used extensively in agricultural and horticultural practices.

They help to initiate rooting in stem cuttings, an application widely used for plant propagation. Auxins promote flowering e.g. in pineapples. They help to prevent fruit and leaf drop at early stages but promote the abscission of older mature leaves and fruits.

In most higher plants, the growing apical bud inhibits the growth of the lateral (axillary) buds, a phenomenon called apical dominance. Removal of shoot tips (decapitation) usually results in the growth of lateral buds. It is widely applied in tea plantations, hedge-making.

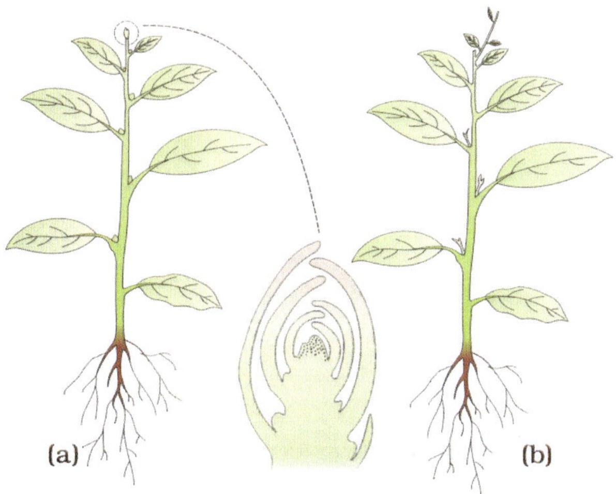

Apical dominance in plants: (a) A plant with apical bud intact, (b) A plant with apical bud removed Note the growth of lateral buds intobranches after decapitation.

Auxins also induce parthenocarpy, e.g., in tomatoes. They are widely used as herbicides. 2, 4-D, widely used to kill dicotyledonous weeds, does not affect mature monocotyledonous plants. It is used to prepare weed-free lawns by gardeners. Auxin also controls xylem differentiation and helps in cell division.

Gibberellins

Gibberellins are another kind of promotery PGR. There are more than 100 gibberellins reported from widely different organisms such as fungi and higher plants. They are denoted as GA_1, GA_2, GA_3 and so on. However, Gibberellic acid (GA_3) was one of the first gibberellins to be discovered and remains the most intensively studied form. All GAs are acidic. They produce a wide range of physiological responses in the plants. Their ability to cause an increase in length of axis is used to increase the length of grapes stalks. Gibberellins, cause fruits like apple to elongate and improve its shape. They also delay senescence. Thus, the fruits can be left on the tree longer so as to extend the market period. GA_3 is used to speed up the malting process in brewing industry.

Sugarcane stores carbohydrate as sugar in their stems. Spraying sugarcane crop with gibberellins increases the length of the stem, thus increasing the yield by as much as 20 tonnes per acre.

Spraying juvenile conifers with GAs hastens the maturity period, thus leading to early seed production. Gibberellins also promotes bolting (internode elongation just prior to flowering) in beet, cabbages and many plants with rosette habit.

Cytokinins

Cytokinins have specific effects on cytokinesis, and were discovered as kinetin (a modified form of adenine, a purine) from the autoclaved herring sperm DNA. Kinetin does not occur naturally in plants. Search for natural substances with cytokinin-like activities led to the isolation of zeatin from corn-kernels and coconut milk. Since the discovery of zeatin, several naturally occurring cytokinins, and some synthetic compounds with cell division promoting activity, have been identified. Natural cytokinins are synthesised in regions where rapid cell division occurs, for example, root apices, developing shoot buds, young fruits etc. It helps to produce new leaves, chloroplasts in leaves, lateral shoot growth and adventitious shoot formation. Cytokinins help overcome the apical dominance. They promote nutrient mobilisation which helps in the delay of leaf senescence.

Ethylene

Ethylene is a simple gaseous PGR. It is synthesised in large amounts by tissues undergoing senescence and ripening fruits. Influences of ethylene on plants include horizontal growth of seedlings, swelling of the axis and apical hook formation in dicot seedlings. Ethylene promotes senescence and abscission of plant organs especially of leaves and flowers. Ethylene is highly effective in fruit ripening. It enhances the respiration rate during ripening of the fruits. This rise in rate of respiration is called respiratory climactic.

Ethylene breaks seed and bud dormancy, initiates germination in peanut seeds, sprouting of potato tubers. Ethylene promotes rapid internode/petiole elongation in deep water rice plants. It helps leaves/ upper parts of the shoot to remain above water. Ethylene also promotes root growth and root hair formation, thus helping the plants to increase their absorption surface.

Ethylene is used to initiate flowering and for synchronising fruit-set in pineapples. It also induces flowering in mango. Since ethylene regulates so many physiological processes, it is one of the most widely used PGR in agriculture. The most widely used compound as source of ethylene is ethephon. Ethephon in an aqueous solution is readily absorbed and transported within the plant and releases ethylene slowly. Ethephon hastens fruit ripening in tomatoes and apples and accelerates abscission in flowers and fruits (thinning of cotton, cherry, walnut). It promotes female flowers in cucumbers thereby increasing the yield.

Abscisic Acid

As mentioned earlier, abscisic acid (ABA) was discovered for its role in regulating abscission and dormancy. But like other PGRs, it also has other wide ranging effects on plant growth and development. It acts as a general plant growth inhibitor and an inhibitor of plant metabolism. ABA inhibits seed germination. ABA stimulates the closure of stomata in the epidermis and increases the tolerance of plants to various kinds of stresses. Therefore, it is also called the stress hormone. ABA plays an important role in seed development, maturation and dormancy. By inducing dormancy, ABA helps seeds to withstand desiccation and other factors unfavourable for growth. In most situations, ABA acts as an antagonist to GAs.

We may summarise that for any and every phase of growth, differentiation and development of plants, one or the other PGR has some role to play. Such roles could be complimentary or antagonistic. These could be individualistic or synergistic.

Similarly, there are a number of events in the life of a plant where more than one PGR interact to affect that event, e.g., dormancy in seeds/ buds, abscission, senescence, apical dominance, etc.

Remember, the role of PGR is of only one kind of intrinsic control. Along with genomic control and extrinsic factors, they play an important role in plant growth and development. Many of the extrinsic factors such as temperature and light, control plant growth and development via PGR. Some of such events could be: vernalisation, flowering, dormancy, seed germination, plant movements, etc.

Photoperiodism

It has been observed that some plants require a periodic exposure to light to induce flowering. It is also seen that such plants are able to measure the duration of exposure to light. For example, some plants require the exposure to light for a period exceeding a well defined critical duration, while others must be exposed to light for a period less than this critical duration before the flowering is initiated in them. The former group of plants are called long day plants while the latter ones are termed short day plants. The critical duration is different for different plants. There are many plants, however, where there is no such correlation between exposure to light duration and induction of flowering response; such plants are called day-neutral plants. It is now also known that not only the duration of light period but that the duration of dark period is also of equal importance. Hence, it can be said that flowering in certain plants depends not only on a combination of light and dark exposures but also their relative durations. This response of plants to periods of day/night is termed photoperiodism. It is also interesting to note that while shoot apices modify themselves into flowering apices prior to flowering, they (i.e., shoot apices of plants) by themselves cannot percieve photoperiods. The site of perception of light/dark duration are the leaves. It has been hypothesised that there is a hormonal substance(s) that is responsible for flowering. This hormonal substance migrates from leaves to shoot apices for inducing flowering only when the plants are exposed to the necessary inductive photoperiod.

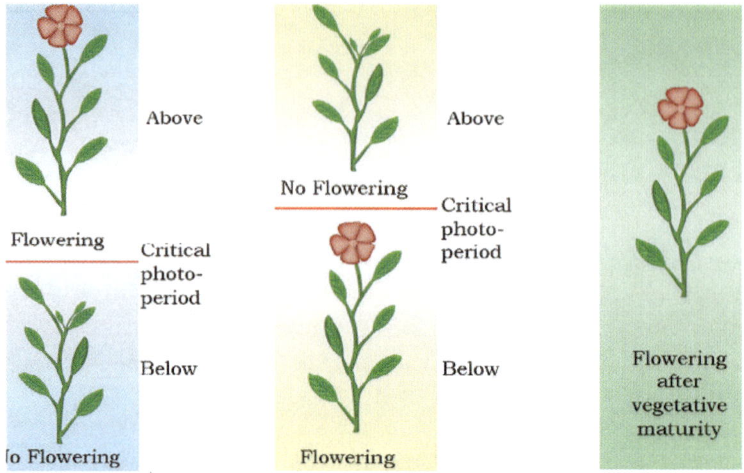

Photoperiodism : Long day, short day and day neutral plants.

Vernalisation

There are plants for which flowering is either quantitatively or qualitatively dependent on exposure to low temperature. This phenomenon is termed vernalisation. It prevents precocious reproductive development late in the growing season, and enables the plant to have sufficient time to reach maturity. Vernalisation refers specially to the promotion of flowering by a period of low temperature. Some important food plants, wheat, barley, rye have two kinds of varieties: winter and spring varieties. The 'spring' variety are normally planted in the spring and come to flower and produce grain before the end of the growing season. Winter varieties, however, if planted in spring would normally fail to flower or produce mature grain within a span of a flowering season. Hence, they are planted in autumn. They germinate, and over winter come out as small seedlings, resume growth in the spring, and are harvested usually around mid-summer.

Another example of vernalisation is seen in biennial plants. Biennials are monocarpic plants that normally flower and die in the second season. Sugarbeet, cabbages, carrots are some of the common biennials. Subjecting the growing of a biennial plant to a cold treatment stimulates a subsequent photoperiodic flowering response.

Plant Ecology

Plants are one of the densest ecotypes with vicariate diversity in their life history depending on their mores. The ecology of land entirely depends upon plants as they are practical pioneers of the planet earth. Initiating from microorganisms to the macroorganism, they meet their living requisites expending plants. Admitting this, different vegetation types, such as grassland

and timberland, are major biomes which help to flux the environmental shifts under global impact. The day-by-day advances in science and technology and ever-emerging needs of living organisms are some anthropogenic activities which had entirely changed the scenario of green systems of the world. Such drastic changes in plant ecology are harmful for all the beings and other respective factors. These substantial amendments occurred due to the earthquakes, industrial effluents, forest fires, carbon ignition, destruction of vegetative and agronomic landforms, etc.

Traditional Ecology

Broadly stating the relationship of plants with the other living organisms and environmental factors is called plant ecology. But elaborating the term, plant ecology is the study of the relationship of plants with the biotic (living organisms such as animals and other plants, bacteria, and fungi) and abiotic factors such as moisture, temperature, sunlight, soil (nutrients and salinity), and water (rainfall and water table) surrounding them. By the passage of time, the addressed issues regarding biosphere came into consideration. Though from the time of Alexander Von Humboldt (father of ecology) the known species of plants were about 20,000, now the number increased up to 40,0000, but the changes in elements of biosphere are increasing the issues such as loss of habitat, plant, animal, microflora, mutation, pollution, and soil sickness. Due to these issues, the most affected living organisms are plants which urged the scientists to investigate the root causes of such drastic changes and commotion to the plant ecology. According to the climate, human and animal interaction, flora, and fauna, the planet earth is categorized into biomes. There are about six major biomes with cutting clarity of subcategories. The largest biome is boreal/coniferous forest; however, the second largest biome is grasslands that are ubiquitous as compared to other biomes; tropical rainforest covers only 6% of the world, but they have the richest biodiversity; however, the hottest biome is desert with the minimal biodiversity; in contrast the coldest biome is alpine forest merely with considerable biodiversity. Specifically, the plant populations have dominantly occupied this globe; according to an estimation, 99.9% area of planet earth is covered with flora. About 350,000 species of plants excluding ferns, bryophytes, and algae are known and documented yet. Among them approximately 20% are under the risk of endangerment. The risk of endangerment or extinction due to natural and unnatural disasters has disturbed the whole food chain and web and is continually pushing toward the worst conditions.

Phylogenetic Ecology

Whenever ecological drift and loss in biodiversity of living organisms are discussed, it is generally apprehended that plants are vanishing due to overgrazing and animals are dying due to the inaccessibility of plants. But this whole globe is alive and functional on a single principle named balanced metabolic dynamics ratio between autotrophs (plants, producers) and heterotrophs (animals and microbes, consumers and decomposer). Basically, this trophic dynamics between producer and consumer within the biosphere is regulated by the transfer of energy from one part of the ecosystem to the other and even within the same ecosystem also known as energy flow in ecosystem. Except solar radiation (external source of energy), all the other energy systems are recycled and balance the dynamics of trophic level followed by complex metabolic mechanisms within the biosphere.

Drafting the origin of plant, their functional types and phylogenetic/evolutionary patterns are the most needed steps to timely track and record the drifts and risks to the ecosystem and biosphere. As the current dynamics, composition and distribution of plants are altered thence, elaborating and redefining the relationship of plants with the factors encompassing them had led the flora and environment on the verge of endangerment are also expatiated, and many successful solutions to indemnify these issues are contributed by the scientists. Sustaining to this several concepts such as phylogeny, phenology, phytosociology, physiology, and anatomy of plants were used for modeling and surveying.

Terrestrial vegetation plays a phenomenal role in management of landscape and hydrological regime. Also the climatic change can be ameliorated by them as they could better regulate biogeological water cycle and sequestrate carbon cycle. The provision of protection against water resources by surface runoff leading toward flood attenuation, aquifer recharging, sea water leveling, water table leveling and fresh water management. Increase in temperature, variation in precipitation, and extreme events have potentially manifested the natural conversational and agricultural management regimes including an indirect risk that was constrained for social and human livelihoods. Although most of the commotion was inflicted due to water regime mismanagement, actually the fraction has been completed by burning practice, by grazing, or by harvesting hay/fodder (directly and indirectly, respectively). Such climatic changes and management conflict of water and vegetation regimes grounded the grasslands toward threats which were also highlighted by Intergovernmental Panel on Climate Change (IPCC) 2014 in their report. Hence, the study of plant ecology is a fountainhead step toward the investigation of cause and solution of biological metabolism and their functioning in the biosphere.

Weed

Weed is a general term for any plant growing where it is not wanted. Ever since humans first attempted the cultivation of plants, they have had to fight the invasion by weeds into areas chosen for crops. Some unwanted plants later were found to have virtues not originally suspected and so were removed from the category of weeds and taken under cultivation. Other cultivated plants, when transplanted to new climates, escaped cultivation and became weeds or invasive species. The category of weeds thus is ever changing, and the term is a relative one.

Weeds interfere with a variety of human activities, and many methods have been developed to suppress or eliminate them. These methods vary with the nature of the weed itself, the means at hand for disposal, and the relation of the method to the environment. Usually for financial and ecological reasons, methods used on a golf course or a public park cannot be applied on rangeland or in the forest. Herbicide chemicals sprayed on a roadside to eliminate unsightly weeds that constitute a fire or traffic hazard are not proper for use on cropland. Mulching, which is used to suppress weeds in a home garden, is not feasible on large farms. Weed control, in any event, has become a highly specialized activity. Universities and agricultural colleges teach courses in weed control, and industry provides the necessary technology. In agriculture, weed control is essential for maintaining high levels of crop production.

The many reasons for controlling weeds become more complex with the increasing development of technology. Plants become weeds as a function of time and place. Tall weeds on roadsides presumably were not problematic prior to the invention of the automobile. However, with cars and increasing numbers of drivers on roads, tall weeds became dangerous, potentially obscuring drivers' visibility, particularly at intersections. Sharp-edged grasses are nominal nuisances in a cow pasture; when the area is converted to a golf course or a public park, they become an actual nuisance. Poison oak (Toxicodendron diversilobum) is rather a pleasant shrub on a sunny hillside in the open country; in a camp ground it is a definite health hazard. Such examples could be given ad infinitum to cover every aspect of agriculture, forestry, highway, waterway and public land management, arboretum, park and golf-course care, and home landscape maintenance.

Weeds compete with crop plants for water, light, and nutrients. Weeds of rangelands and pastures may be unpalatable to animals, or even poisonous; they may cause injuries, as with lodging of foxtails (Alopecurus species) in horses' mouths; they may lower values of animal products, as in the cases of cocklebur s (Xanthium species) in wool; they may add to the burden of animal care, as when horses graze in sticky tarweeds (Madia species). Many weeds are hosts of plant disease organisms. Examples are prickly lettuce (Lactuca scariola) and sow thistle (Sonchus species) that serve as hosts for downy mildew; wild mustards (Brassica species) that host clubroot of cabbage; and saltbrush (Atriplex species) and Russian thistle, in which curly top virus overwinters, to be carried to sugar beets by leafhoppers. Many weeds are hosts of insect pests, and a number are invasive species.

Characteristics of Weeds

Certain characteristics are associated with and allow the survival of weeds. Weeds posses one or more of the following:

- Abundant seed production,
- Rapid population establishment,
- Seed dormancy,
- Long-term survival of buried seed
- Adaptation for spread,
- Presence of vegetative reproductive structures,
- Ability to occupy sites disturbed by human activities.

There are approximately 250,000 species of plants worldwide; of those, about 3% or 8000 species behave as weeds.

Weeds are troublesome in many ways. Primarily, they reduce crop yield by competing for water, light, soil nutrients, and space. Other problems associated with weeds in agriculture include:

- Reduced crop quality by contaminating the commodity,
- Interference with harvest,

- Serve as hosts for crop diseases or provide shelter for insects to overwinter,
- Limit the choice of crop rotation sequences and cultural practices,
- Production of chemical substances which are toxic to crop plants (allelopathy), animals, or humans.

Costs of Weeds

Weeds are common on all 485 million acres of U.S. cropland and almost one billion acres of range and pasture. Since weeds are so common, people generally do not understand their economic impact on crop losses and control costs. In 1991, the estimated average annual monetary loss caused by weeds with current control strategies in the 46 crops grown in the United States was $4.1 billion. If herbicides were not used, this loss was estimated to be $19.6 billion. Losses in field crops accounted for 82% of this total.

Another source estimates that U.S. farmers annually spend $3.6 billion on chemical weed control and $2.6 billion for cultural and other methods of control. The total cost of weeds in the United States could approach $15 to $20 billion dollars. Also, weed control and other input costs (e.g., seed, fertilizer, other pesticides, fuel) vary with the crop. For example, in the mid-90s, herbicides for soybeans cost $30/acre or about 47% of the $63/acre in total purchased input. For corn, the cost was $32/acre or about 28% of the $114/acre in total purchased input. And for wheat it was $6 or about 6% of the total $96/acre inputs. Several factors help determine the relative costs of herbicides from one crop to another and include the competitive ability of the crop, the weeds present, the contribution of non-chemical control practices, the tillage method, management decisions, and the value of the crop.

Benefits of Weeds

Despite the negative impacts of weeds, some plants usually thought of as weeds may actually provide some benefits. Some attributes include:

- Soil stabilization;
- Habitat and feed for wildlife,
- Nectar for bees;
- Aesthetic qualities;
- Add organic matter;
- Provide genetic reservoir;
- Human consumption; and
- Provide employment opportunities.

Weeds have a controversial nature. But to the agriculturist, they are plants that need to be controlled, in an economical and practical way, in order to produce food, feed, and fiber for humans and animals. In this context, the negative impacts of weeds indirectly affect all living beings.

Modern weed control can be classified as mechanical, chemical, or biological.

Mechanical Control

Peanut farming

Mechanical weed control began when humans first pulled weeds from their cereal crops and attempted to grow single plant species, free from all plant competition. That was the start of monoculture, a method that since has come to dominate agriculture. Contrary to the principles of ecology, farmers throughout the world grow the major food, fibre, and forage crops in a monoculture because experience has shown that the highly improved modern crop species give their highest yield under this system.

From hand pulling, humans devised simple tools such as the spud, the knife, and the hoe to eliminate weeds. For thousands of years, from the Egyptian culture to the Renaissance, those simple methods were used. The first efforts to turn away from simple hand methods and mechanize the arduous task of weed control began in 17th-century England. Since then there has been continuous improvement of agricultural tools used to destroy weeds and of cultural methods employed to minimize weed growth. The principal virtue of cultivation of row crops is the control of weeds. Any method of weed control that minimizes tillage tends to conserve soil structure and maintain fertility.

Tillage: Farm tractor tilling row crops as a form of mechanical weed control.

In addition to tillage, other mechanical methods of weed control involve burning, grazing, and the use of ducks or geese in certain crops (in cotton and mint especially). All of those methods have drawbacks: there is the arduous, painful nature of hand weeding; the repetitious and often harmful nature of clean tillage with machinery; the slow, fuel-consuming nature of burning; and the costly requirement of livestock or fowl for the biological grazing methods. Tillage, still the most widely used method of row-crop weed control, has been greatly improved by development of precision seeding and close preadjustment of tiller tools, allowing the passage of weed knives within an inch or less of the young crop plants. Despite these improvements it is known that weed knives injure crop roots, especially late in the tillage season. Additionally, tillage tools can spread perennial weeds rapidly, bringing about rapid infestation of whole fields.

Such methods as crop rotation, use of smother crops, use of weedfree seed, mulching and covering, and cleaning of machinery to prevent spread of weed seeds are also classified as mechanical.

Chemical Control

Chemical weed control has been used for a very long time: sea salt, industrial by-products, and oils were first employed. Selective control of broad-leaved weeds in fields of cereal crops was discovered in France in the late 1800s, and this practice soon spread throughout Europe. Sulfates and nitrates of copper and iron were used; sulfuric acid proved even more effective. Application was by spraying. Soon sodium arsenite became popular both as a spray and as a soil sterilant. On thousands of miles of railroad right-of-way, and in sugarcane and rubber plantations in the tropics, the hazardous material was used in tremendous quantities, often resulting in the poisoning of animals and occasionally humans. Diesel oil, as a general herbicide, and sodium dinitrocresylate (Sinox), as a selective plant killer, were introduced during the first three decades of the 20th century.

Herbicide

Sinox, the first major organic chemical herbicide, was developed in France in 1896. In the late 1940s new herbicides were developed out of the research during World War II, and the era of the miracle weed killers began. Within 20 years over 100 new chemicals were synthesized, developed, and put into use. Chemical weed control superseded both plant-disease and insect-pest control in economic impact. The year 1945 marked the beginning of a new era in chemical weed control. Introduced then were 2,4-D (2,4-dichlorophenoxyacetic acid), 2,4,5-T (2,4,5-trichlorophenoxyacetic acid), and IPC (isopropyl-N-phenylcarbamate), the first two selective as foliar sprays against broad-leaved weeds, the third selective against grass species when applied through the soil. The new herbicides were revolutionary in that their high toxicity allowed for effective weed control at dosage rates as low as one to two pounds per acre. This contrasted with carbon bisulfide, borax, and arsenic trioxide, which were required at rates of up to one ton per acre, and with sodium chlorate, required at rates of around 100 pounds per acre. However, some of those early herbicides, including 2,4,5-T, were later deemed unsafe for humans and the environment and were discontinued in many countries. Effective herbicides have continued to be developed, and some, such as glyphosate, are widely used around the world.

Herbicides may be grouped into two categories: selective and nonselective. Each category may be subdivided into foliage-applied and soil-applied materials and, in cases where field crops are treated, the application may be made before sowing the crop (preplanting), after sowing but before emergence of seedlings (pre-emergence), or after seedlings have emerged (postemergence).

A great advantage of chemical over mechanical weed control is the ease of application. This is particularly true in cereal croplands, pastures, rangelands, forests, and other situations where an airplane can be used. Many millions of acres are treated from the air each year, and modern equipment for treating row-crop land with herbicides has made weed control increasingly convenient. Sprayers, soil incorporation equipment, and spreaders for pelleted herbicides have all added to the convenience of, and removed uncertainty from, herbicide application. Machinery is available that simultaneously builds up beds, plants the seed, sprays with insecticide, and incorporates fertilizer and pre-emergence herbicide all in one operation.

Certain agricultural plants, known as herbicide-resistant crops (HRCs), have been genetically engineered for resistance to specific chemical herbicides, notably glyphosate. These genetically modified organisms (GMOs) have been available since the mid-1980s and enable effective chemical control of weeds, since only the HRC plants can survive in fields treated with the corresponding herbicide. However, because these crops encourage increased application of chemicals to the soil rather than decreased application, they remain controversial with regard to their environmental impact and general safety.

Biological Control

Efforts to eradicate weeds through biological control are a fairly recent development. An early report from 1902 described the importation of insects from Mexico to Hawaii in an effort to control Lantana, an imported shrubby climbing weed that had spread over thousands of acres of pastureland, rendering them useless for grazing. Prickly pear cacti have been very effectively controlled in Australia; some 24 million hectares (60 million acres) have been converted from cactus thicket to plowland and pasture by the cactus moth (Cactoblastis cactorum), which was introduced from Argentina in 1925. By 1933 the major cactus areas were under control.

Another successful use of biological weed control was in the western United States, where Saint-John's-wort, or Klamath weed (Hypericum perforatum), was subjected to depredation by three insect species, beginning in California in 1945. The release of two insects of the genus Chrysolina and one of the genus Agrilus continued for a number of years, and the effort was carried to Oregon, Washington, and Idaho by 1950. The insects spread rapidly after introduction and became established, ultimately reclaiming some 1 million hectares (2.5 million acres) of rangeland from the invasive plant.

A number of vertebrate animals have been used to control certain specific weeds. Sheep and goats have been employed to control brushy plants on rangelands in many countries. Their effectiveness is evident in parts of the Middle East and Africa where dry range and desert lands have been almost completely denuded by grazing goats. In these cases, however, the destructiveness of the goats far outweighs their usefulness in plant pest control, which indicates the need for rational management in all efforts at weed control.

Geese have been used to control weeds in cotton fields in California and in mint plantations in Oregon. Certain fishes are useful in keeping aquatic plants under control; examples are the Congo tilapia, the Israeli carp, and the grass carp. The Florida manatee is known to consume many aquatic plants, and the giant ramshorn snail (Marisa cornuarietis) feeds on alligator weed, pondweed, and water hyacinth.

Although much more desirable than chemical weed control from the standpoint of longtime ecological effects, biological weed control has definite limitations. It is ideal for situations where a weedy plant was introduced free of its natural predators, and there are places where it offers the only hope for coping with serious weed situations. However, biological control using insects is limited almost entirely to perennial plants and is restricted mainly to weeds of uncultivated areas. Given the broad spectrum of agricultural weed species, the great number of seeds in the soil seed bank, and the fact that many weed seeds can survive for decades in the soil, success against annual crop weeds with biological agents is not very feasible. In those cases cheap and effective herbicides have proved most useful.

Additionally, biological weed control tends to be only periodically effective. Weed populations may be reduced initially to very low levels by the control insects; however, they recover when the insects die off for lack of food. The insect predators may then flourish again in reciprocal cycles, but complete control may never be achieved.

Finally, the introduction of alien organisms as biological control is hazardous in that these same organisms may become pests in the new habitat. Kikuyu grass, which was introduced into California to prevent soil erosion on hillsides and roadways, soon spread into orchards, turf, and crop areas, where it became a serious weed.

Weed Control Tips

Never Let them Set Seed

This is the #1 rule with weeds. Some varieties produce tens of thousands of seeds from a single plant, multiplying your weed control problems for years to come. So make certain you remove weeds around your home before they flower and produce seeds. Pigweed, purslane, Shepherd's purse, chickweed, and lambsquarters are examples. Their seeds are very small and light enough to be blown by the wind over short distances, spreading profusely and often surviving for decades in the soil.

Remember that it may take a few years to get weed-free. Seeds of most annual weedy grasses die after two or three years, but some broadleaf weed seeds can last much longer. On average, though, the bulk of your weed seeds will be depleted in about five years if no additional seeds are added. That means diligence is the key.

Mulch

For further weed suppression throughout the growing season, apply two to three inches of mulch or use landscape fabric or black plastic. Mulch not only blocks weed seeds from sunlight so they do not germinate, but also promotes better water retention, provides needed nutrients as it decomposes over time, and moderates soil temperatures. If you mow or blow leaves in the fall, be sure to get a shredder to turn those leaves into garden mulch—and save yourself the costs of buying bags of mulch.

Turn to Tools

A garden hoe, tiller or even hand-weeding can work, especially if the space you're tending is fairly small. And keep your tools (garden hoe, spade, mower, tiller) clean to keep from spreading weed seeds or plant parts that you encounter. Tillers are ideal for aerating soil and keeping flower and plant beds weed-free, turning what can be a difficult job into easy, productive work.

Establish a Perimeter

Pay special attention to the area adjoining your flower bed, garden, natural area or lawn and establish a weed-free perimeter. Mow or mulch the area or pull or dig up weeds as they emerge. You'll help to reduce the number of new weed seeds in the area you want to protect. Also, a good trimmer can make it easier to reach weeds along garden beds, posts, and tight spots.

Pay Special Attention to Perennial Weeds

When you deal with perennials, you need dig up any roots, underground tubers and rhizomes without leaving fragments behind. New weeds can grow from any pieces that break off and remain in the soil. It does also help to cut off the emerged green part of the weed with your hoe or mower—repeating the process quickly each time it regrows. Without leaves needed for photosynthesis, the underground plant parts will become weakened and may eventually die.

What makes a Plant a Weed

Weeds can be defined in a variety of ways: a plant growing where it isn't wanted, a plant that interferes with farming or grazing, a plant that was not intentionally sown, a plant that is persistent and detrimental to the plants around it, among others. Basically, what it comes down to is that weeds are plants that humans don't want to have around for one reason or another.

In the United States, there is a federal weed list and many states have their own weed lists, too. Some common flowers appear on the state weed lists yet appear in gardens and flower bouquets across the country. For instance, baby's breath (Gypsophila paniculata) is considered a noxious weed in California, while the ox-eye daisy (Chrysanthemum leucanthemum), your typical white daisy with the golden centre, is considered a weed in numerous states, including Colorado, Kansas, Ohio, and West Virginia.

Some weeds are simply plants that are not wanted in the place where they are growing, but are useful at other times. For instance, a wild raspberry plant smothering a crop of strawberries planted by a farmer. In other areas, a weed might be an invasive species of plant which threatens a country's natural flora. An example of this is broom, or Cytisus scoparius, which grows abundantly in

Scotland and is used in ornamental landscaping, sand dune stabilization, and wasteland reclamation. It was even used as a symbol of the Plantagenet kings. However, in Australia, New Zealand, and parts of North America, it is considered a pest species and often eradicated.

Sometimes weeds were considered undesirable in the Old World before they came to the new one. Weed seeds were unintentionally spread after they were mixed in with seeds of other plants, like those of vegetables. Other times, farmers found weeds useful to help other plants grow—such as sweet briar in New Zealand, which was used to help grafted roses grow, and then became a pest when it spread.

Typically, weeds also possess certain characteristics that make them particularly irritating to the human race: they grow and spread very quickly, which can hurt the surrounding plants by depriving them of sunshine and nutrients in the soil. Their spread can be attributed to their abundant seed production and the hardiness of those seeds, which can lay dormant in the soil for long periods of time. Weeds themselves are also hardy, often able to live in extreme conditions and in places affected by human interference. To sum it up, weeds are typically extremely difficult to exterminate from a garden bed entirely.

Other than acting as an invasive species in new areas, weeds also are often considered such as they are plants that pose a threat to farmers and their crops. Weeds can contaminate crops, reducing crop quality; they can host diseases which kill the crop; shelter insects in the winter which will then eat the crop; and they could also produce toxins, infecting the crop with chemicals that are dangerous for human consumption.

Needless to say, farmers tend to want to keep weeds away from their crops. They spend billions of dollars every year on herbicides to keep weeds at bay, but even that doesn't stop all of them. Additional money is lost on crops that fail or are contaminated because of weeds. Their loss in profit means higher prices at the grocery store for the consumer.

Despite all of the flaws of weeds, many do have some benefits, too. They provide soil stabilization and nectar for bees. They can be food for wildlife or even humans, as well as occasionally used for medicinal purposes. And, of course, as alluded to, weeds are big business and many jobs are created because of them. After all, how many fewer workers would be needed and fewer home lawn and garden products if weeds were as easy to kill as that houseplant you're always struggling to keep alive.

Weed Biology and Ecology

Weed Biology

Most plants have one of three main life cycles—annual, bien- nial, or perennial. An annual plant completes its life cycle in one year as it germinates, grows, flowers, sets seed, and dies. Most of the weeds in agricultural fields are annuals such as pigweeds and foxtails. Most Crops are also annuals.

A biennial is a plant that needs two growing seasons to complete its life cycle. The first year, biennials produce vegetative growth in the form of a rosette where all the leaves come from the center crown. Biennials go dormant over the winter and in the second year, regrow, flower, set seed, and die. Some common biennials are musk thistle and mullein.

Musk thistle is a biennial weed.

The rosette of musk thistle.

A perennial is a plant that lives for three or more years as it grows, flowers, and sets seed in a continuous cycle over several seasons. Canada thistle and quackgrass are perennials. Additionally, perennials have special underground parts (rhizomes, tubers, stolons) that allow them to spread vegetatively as well as by seed.

Reproduction in Weeds

Plants have two main modes of reproduction, by seed or vegetatively. Most annuals and bienni- als reproduce by seed, and in the case of weeds, the production is often quite prolific. For example, redroot pigweed can produce over 100,000 seeds/plant. Perennials can reproduce by seed as well as by vegetatively via rhizomes and stolons. A rhizome is an underground stem that sends out roots and shoots from its nodes. A stolon is an aboveground stem that grows from an existing stem at a node, like a strawberry runner.

Amount of seed produced per plant by different weed and crop species.

	Species	Seeds/Plant		Species	Seeds/Plant
Weed	Canada thistle	680/stem	Weed	Smartweed	19,500
	Giant foxtail	900		Waterhemp	23,000
	Cocklebur	900		Common chickweed	25,000

	Wild mustard	1,200		Burdock	31,600
	Wild buckwheat	1,200		Shepardspurse	38,500
	Common ragweed	3,500		Common purslane	52,300
	Yellow foxtail	6,500		Lambsquarters	72,500
	Common sunflower	7,200		Redroot pigweed	117,400
	Velvetleaf	7,800		Horseweed (marestail)	200,000
	Eastern black nightshade	10,000		Common mullein	223,200
	Giant ragweed	10,300			
	Hemp dogbane	12,000	Crop	Corn	800
	Kochia	14,600		Soybean	50
	Dandelion	15,000		Winter wheat	110

A tuber is a thickened part of a rhizome or stolon that is used as a place of storage for starch (e.g. Jerusalem artichoke, yellow nutsedge). Many plants that have above ground stolons also form horizontal, below ground rhizomes.

Seeds, rhizomes, stolons, and tubers are all considered propagules because they are able to generate entire new plants. Weeds potentially produce very many propagules per plant, but actual productivity is much lower in competition with the crop or at high weed densities. The crop-weed interaction can reduce potential weed seed production dramatically, as much as 50 percent.

Reducing risk: life cycles nd reproduction. rease weed risk by identifying the plant life cycle and reproduction mode of your problem weed species. For example, annuals can be contained through tillage or mowing prior to seed production. On the other hand, tillage can increase a perennial by breaking up the roots and creating new plants more quickly.

Weed Ecology

Weed ecology is the study of the interaction or relationship between a weed and its environment (other living organisms as well as abiotic factors). Ecology is concerned with growth characteristics and adaptations that enable weeds to survive the change in the environment. Man plays an important role in changing the environment by altering the crop husbandry practices and by maintaining weed free monocrop or multicrop culture. For effective weed control, the study on both biology and ecology of a weed species are important.

Weed Seedbanks

It is hard to imagine the number and variety of weed seeds in a field. Once weed has produced seed and dispersed them in the soil, the majority of the seeds remain for a long period of time. This reservoir of viable seeds in the soil is called a seed bank. If those weeds are allowed to grow and go to seed, an ugly cycle of weed seed replenishment can frustrate even the most attentive farmer. In any given year, only a small percentage of seeds in the seed bank germinate due to a variety of seed dormancy mechanisms. The rest of those seeds remain waiting for the next opportunity to grow.

Number of viable weed seeds in four agricultural fields in Minnesota. Soil was sampled to a depth of 6 inches.			
Location	County	Seed/ft^2	Seed/Acre (in millions)
Sacred Heart	Renville	118	5.1
Danube	Renville	184	8.0
Morris	Stevens	586	25.5
Waseca	Waseca	7661	333.7

Percent reduction of weed seed production when weeds emerge after crop emergence as compared to when weeds emerge with crop. The amount of seed is dramatically reduced when weeds emerge after the crop.			
Weed	Crop	Weed Emergence (# Weeks after Crop)	% Weed Seed Reduction
Waterhemp	Corn	3	95
Waterhemp	Soybean	3	81
Giant ragweed	Corn	6	99
Giant ragweed	Soybean	6	78
Velvetleaf	Corn	3	60

A critical aspect of weed management is reducing weed seed production. Crop competition can reduce potential weed seed production. Thus, weed seedbanks can be decreased in response to good management, while seedbank increases will occur in years with poor weed management. Producers should remember that prevention is better than finding a cure.

Reducing risk: weed seedbanks. Practice good weed management on the whole farm to prevent increases in weed seedbanks. Prevent weeds from going to seed as much as possible. Clean tillage equipment to prevent movement of underground reproductive structures.

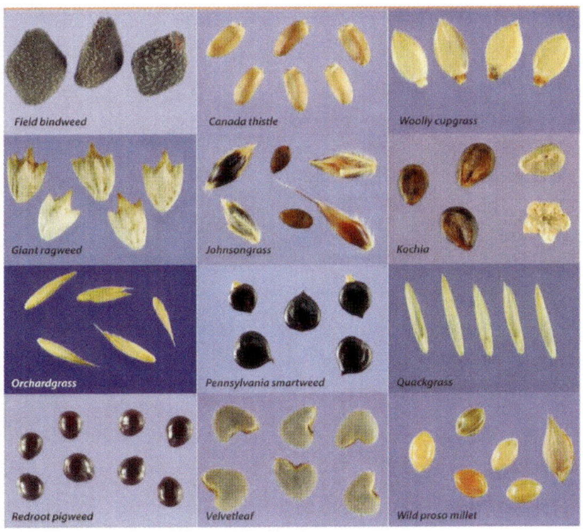

Weed seeds have a variety of sizes, shapes, and colors. Seeds of 12 weed species are shown. Field bindweed, Canada thistle, giant ragweed, johnsongrass, kochia, orchardgrass, Pennsylvania smartweed, quack- grass, redroot pigweed, velvetleaf, wild proso millet, and woolly cupgrass.

Weed Dispersal

Most agricultural weeds (~75 percent) lack any obvious dispersal mechanisms and fall close to the parent plant. But weeds do move around, and dispersal mechanisms are as varied as the number of weed species. Weed seed dispersed by wind (e.g. dandelion, thistles) usually has structural modifications making them very lightweight in the air. Flooding and irrigation are good dispersal mechanisms as most seeds can float and can live in the water for some time. Birds and animals can move seed great distances. Seed contamination via weed mimicry (e.g. clover in alfalfa) is also a source of dispersing weed seeds to new sites. Agricultural activities like planting contaminated crop seed, using unclean harvest equipment and tillage equipment, and moving machinery between fields are significant weed seed dispersal procedures. Spreading manure is another common way to disperse weed seed. Composting manure can eliminate some weeds. Knowing the potential sources of weed contamination and cleaning equipment are good starting points to reducing new infestations and lowering farmer risk.

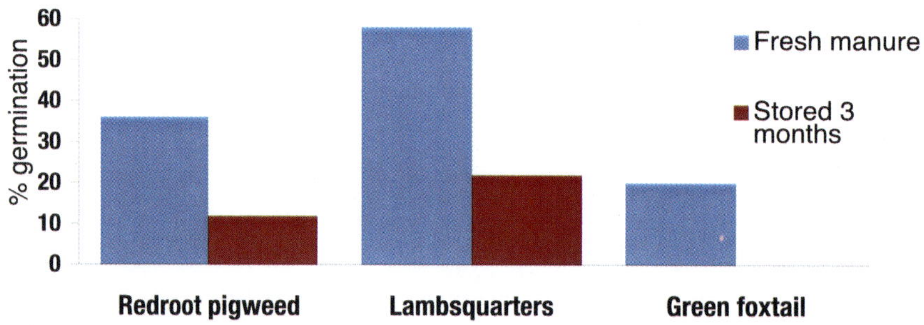

Percent germination of three weeds in fresh manureand manure that has been stored for three months. Green foxtail had zero percent germination after three months. Weed seed can still remain viable after livestock digestion and even after storage.

Scale of distance of weed seed dispersal mechanisms. Dispersal can be as a result of human activity (irrigation) or as a result of natural activity (wind).

Dispersal Mechanism	Within Fields	Between Fields	Between Regions
Livestock (transported)		✓	✓
Contaminated seed		✓	✓
Irrigation water		✓	
Manure		✓	
Combines	✓	✓	
Livestock (walking)	✓	✓	
Birds	✓	✓	
Plows	✓	✓	
Wind	✓	✓	
Insects	✓		
Rain	✓		

Reducing risk: weed dispersal. Be aware of the routes of dispersion. Always start with clean, weedfree seed or certified seed. Inspect and clean machinery. When using compost instead of manure, ensure it has been properly composted to kill as many weed seeds as possible.

Dormancy

Weed seed dormancy is another type of dispersal—dispersal through time instead of space. When seed is dispersed, most does not immediately germinate. It remains dormant in a sort of sleeping stage until conditions are right. The factors that break dormancy are unpredictable and dependent on the species, the weather conditions, even physiological factors within the seed itself. Over time seeds that do not germinate go from dormant to non-viable (dead). Weed persistence in the seedbank will vary among species. Again, this is species and climate condition dependent but can be further manipulated by farmers who have identified their weed problems and are proactive about crop rotation and weed seed burial via tillage.

Weed and crop seed persistence in soil. The approximate number of years it takes to reduce weed seed populations by 50 and 99 percent.			
Broadleaves	Lambsquarters	12	78
	Velvetleaf	8	56
	Cocklebur	6	37
	Pennsylvania smartweed	4	26
	Redroot pigweed	3	20
	Shepardspurse	3	19
	Curly dock	3	17
	Waterhemp	2	16
	Common ragweed	1.5	10
	Wild mustard	1	7
	Common sunflower	0.5	2
	Hemp dogbane	0.5	2
	Giant ragweed	0.5	2
	Kochia	0.5	2
Grasses	Yellow foxtail	5	30
	Barnyardgrass	2	10
	Large crabgrass	1.5	8
	Giant foxtail	1	5
Crops	Wheat	1	2
	Canola	2	4
	Soybean	1	2
	Corn	2	4

Seed Characteristics

Weed seeds have general characteristics that producers can use to manage them. Here are some general rules:

- Seed of broadleaves are more persistent in the soil compared to grasses.
- Annuals and non-rhizomatous perennials tend to be persistent in seed banks.
- Small, round seeds tend to be more persistent than large or elongated ones.
- Small seeds are more likely to go dormant immediately.
- Large seeds are less susceptible to allelopathic compounds such as from a rye crop.
- Small seeds do not emerge well from depths greater than two inches.

Seed size and depths at which inhibition of seed germination or emergence occurs. There are depths at which weed seed will not be able to emerge, usually corresponding to seed size.			
	(mm)	(in.)	(in.)
Common purslane	0.6	1.5	3.1
Common chickweed	1.0	1.4	3.1
Redroot pigweed	1.0	2.1	3.9
Wild mustard	1.5	1.7	3.9
Lambsquarters	1.5	1.9	3.9
Black nightshade	1.6	2.1	3.9
Prostrate knotweed	2.0	2.1	3.9
Large crabgrass	2.5	1.6	3.1
Jimsonweed	2.5	2.4	4.7
Canada thistle	3.0	2.1	3.9
Velvetleaf	3.0	2.8	4.7
Barnyardgrass	3.5	2.1	3.9
Johnsongrass	4.0	2.5	4.7
Field bindweed	5.0	2.7	4.7

Producers can use traits such as persistence and germination depths of different weeds as a guide to the effectiveness of burying weed seed with tillage. Thus, shallow cultivation will keep seeds on top and reduces germination by not providing them with conditions like adequate moisture that encourage germination. Deep cultivation will bury large seeds like cocklebur. Large seeds are less persistent and if buried deep enough, they will not survive. However, small weed seed survival is increased by burial, as they will go dormant until conditions bring them back to the surface.

Reducing risk: dormancy and seeds. Be aware that some field operations will expose weeds to conditions that break seed dormancy. Viable buried seed that is brought to the surface via deep tillage may germinate. Reduced or shallow tillage may leave dormant seeds buried, preventing germination, but can also leave small seeds closer to the surface, providing them greater opportunity to germinate.

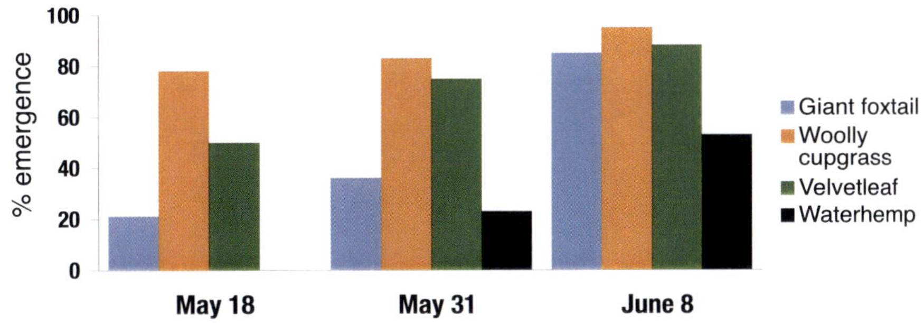

Percent emergence by date for four weeds in Ames, IA. In this example, weeds like giant foxtail, woolly cupgrass, and velvetleaf will mostly be emerged by June 8th, while the waterhemp population is only halfway finished.

Weed Emergence

Weeds rarely emerge in a single uniform flush. Emergence for each weed species is based on a wide variety of factors depending on the weather, soil type, tillage system, prior crop, and crop rotation. But year to year emergence, and the duration of emergence, of a known species is fairly consistent. Some weeds emerge over a span of two to three weeks (giant ragweed and woolly cupgrass), four to seven weeks (lambsquarters, common ragweed, and yellow foxtail), and others over a more prolonged eight to ten weeks (velvetleaf, giant foxtail, and waterhemp). A variety of computer tools, usually based on soil type, growing degree days, and tillage are available to farmers.

Reducing risk: emergence. Be able to identify weed seedlings on your farm. Know the timing and emergence of weeds to synchronize mechanical weed control operations.

Relative emergence of weeds in Minnesota								
Group 0		Group 1	Group 2	Group 3	Group 4	Group 5	Group 6	Group 7
Weed species	Horseweed Downy brome Field pennycress Sheards-purse Biennial thistles	Foxtail barley Kochia Prostrate knotweed Wild mustard Russian thistle White cockle	Quackgrass Orchardgrass Giant ragweed Pennsylvania smartweed Lambsquarters Wild oats Hairy nightshade Common sunflower	Smooth brome Common ragweed Woolly cupgrass Velvetleaf Wild buckwheat	Canada thistle Giant foxtail Yellow nutsedge Redroot pigweed Cocklebur	Green foxtail Common milkweed Hemp dogbane Barnyard grass Yellow foxtail Wild proso millet	Black nightshade Shattercane Venice mallow Waterhemp Jerusalem artichoke	Fall panicum Crabgrass Jimson-weed Witchgrass

Approx. date in Minnesota	Fall	April	Early May	Early to mid May	Mid to late May	Late May/ early June	Early to mid-June	After mid-June
Concurrent Field Operations	Fall tillage	Spring tillage Seedbed prep Small grains planting	Seedbed prep	Pre-emergent weed control Corn planting Seedbed prep	Pre-emergent weedcontrol Corn planting Soybean planting	Post-emergent weed control Soybean planting Alternative crop planting	Post-emergent weed control Cultivation	Cultivation

Weed Seed Fate and Seedling Mortality

Like all seeds, a weed seed's fate in a field is no mystery. It can germinate and live, be removed by wind or water, germinate and die, decay over time, become inviable (dead), stay dormant, or get eaten! Weed seed mortality is derived in three main ways: seed predation in the soil, aging of the seed over time, and germination at the wrong depth or time of year. The ultimate fate of a weed seed will vary by species.

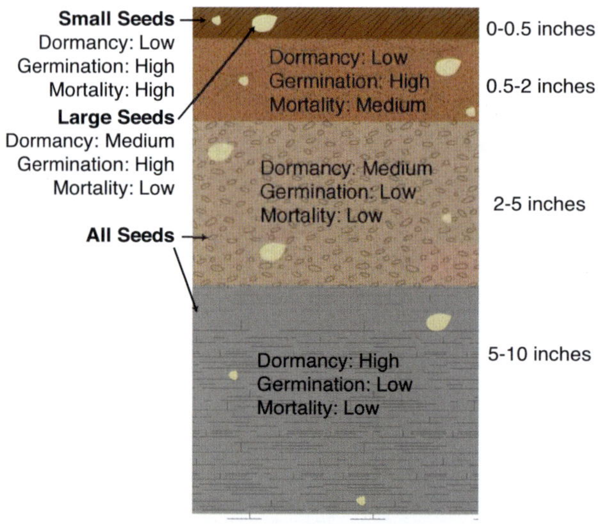

Weed seed fate depends on placement in the soil profile. Once seeds are past 1/2-inch soil depth, fates are similar regardless of size.

Weed Management Decision Tools

One of the most important decisions that organic producers make is when to time weed control operations for effective results. Knowing when the weeds will be present and when they will most easily be controlled is an integral part of this decision. There are several weed software programs that can aid in the decision-making process. WeedCast is an example useful for producers in the Midwest. Weather and site data are entered by a producer and emergence information about particular weeds in their fields are displayed.

But for seeds that do germinate and live, weed seedling survival after emergence is very high. Rates of natural mortality due to disease, herbivory and drought are low for established weeds in annual crops. So, if a weed makes it to seedling stage, its rate of survival to maturity is

25-75 percent, up to even 90 percent. Mortality also decreases with increasing plant size and age. Despite starting small, weed seedlings quickly catch up with crop seedlings—they like the same growing conditions as the crop seed does. Weed seedlings have a very high relative growth rate (amount of growth/biomass) and quickly establish a fine root network for nutrient uptake. Smaller seeds have small reserves compared to crops, making them more dependent on soil nutrients.

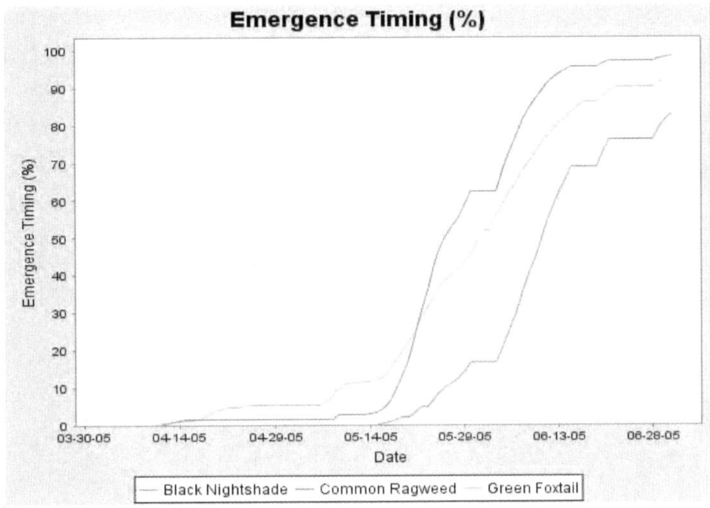

Example output from WeedCast showing emergence timing for black nightshade, common ragweed, and green foxtail.

Weed Density

Weed density is a function of the weed seedbank and its emergence rate. The density of a weed cohort has several consequences. Farmer risk with respect to weed density includes yield loss and problems of future weed management. It is worth noting that density, at least at initial germination, may not be indicative of later plant densities, as some plants will die due to crowding, crop competition, and various climate factors.

Weed Effects on Crops

The negative effects of weeds are well-known. The level of damage to a crop will be dependent on factors relating to weeds such as species present and weed density, but the crop itself will also be a factor. Both the weeds and crop are considered when determining the weed thresholds where management options should be considered.

Weed Thresholds

While weeds may not be wanted, how many are too many? Total eradication, while possible, could be excessively expensive, incur unacceptable environmental damage, and deprive the farmer of some of the ecological services—actual benefits—of having unwanted plants on the farm.

Weed seed bank densities and seedling emergence in row crops in Morris, MN.

Densities will be dependent on the weed species and the initial weed populations in the seed bank.

	Green foxtail	Redroot pigweed	Lambsquarter	Other weeds
Seeds in seedbank at start of season	972	672	379	59
Seedlings prior to crop planting	16	0	6	4
Seedlings after crop planting	43	10	8	4
Seedlings after interrow cultivation	13	4	7	1

Weed Benefits

It may be difficult to imagine, but weeds can provide ecological benefits. If seed production can be prevented, producers may be able to take advantage of some of these benefits.

- Protect against soil erosion.
- Fix nitrogen (if weed is a legume).
- Add organic matter.
- Provide habitat for beneficial organisms.
- Conserve soil moisture.
- Scavenge nutrients.
- Contribute forage.
- Increase biodiversity.

A weed threshold is the number of weeds it takes before a producer deems them necessary to control.

In developing thresholds, the number and timing of weed control operations need to be balanced against minimizing crop injury, soil damage, and costs. Good yields rely on the relative timing of emergence of crop versus weeds, the time it takes for the crop to reach a good height over the weeds, and how rapidly the canopy of the crop closes.

Weed thresholds are one of two main categories—competitive or economic. Competitive thresholds are the levels at which weeds negatively affect yield. They are determined by weed density, duration of interference, and crop reduction. Crops are not equal in their ability to compete with weeds, and weeds vary in their ability to compete with the crop. Often, if more than one weed species is present, the competitive effects are not additive. As weed density increases, weeds compete with the crop and each other— making it hard to predict yield loss. Crops can tolerate weeds up to a point—but a critical period arrives at which weeds must be managed to avoid crop loss. Critical periods vary between crops.

Economic thresholds examine the value of the management decision—at what point is the cost of management worth the amount of yield gain? Economic thresholds are more difficult to estimate as they must account for a given crop, weed community, cost of management, commodity price, and amount of potential yield loss.

Risk levels of weed species on corn and soybean yield.			
Broadleaf Weeds	Risk	Grass Weeds	Risk
Giant ragweed	High	Johnsongrass	Moderate
Common sunflower	High	Quackgrass	Moderate
Common cocklebur	High	Barnyardgrass	Low
Velvetleaf	High	Giant foxtail	Low
Lambsquarters	High	Green foxtail	Low
Common ragweed	High	Yellow foxtail	Low
Jimsonweed	High	Large crabgrass	Low
Common waterhemp	Moderate	Fall panicum	Low
Redroot pigweed	Moderate	Witchgrass	Low
Kochia	Moderate		
Pennsylvania smartweed	Moderate		
Canada thistle	Moderate		
Field bindweed	Low		
Horseweed	Low		
Eastern black nightshade	Low		

Reducing risk: weed thresholds. Be observant of weeds levels and yields for your farm to develop an idea of weed thresholds for individual situations. Good record keeping will be helpful. When weed thresholds are met, apply appropriate measures. Realize that there will be times when weeds may not need to be controlled.

Crop Competitiveness

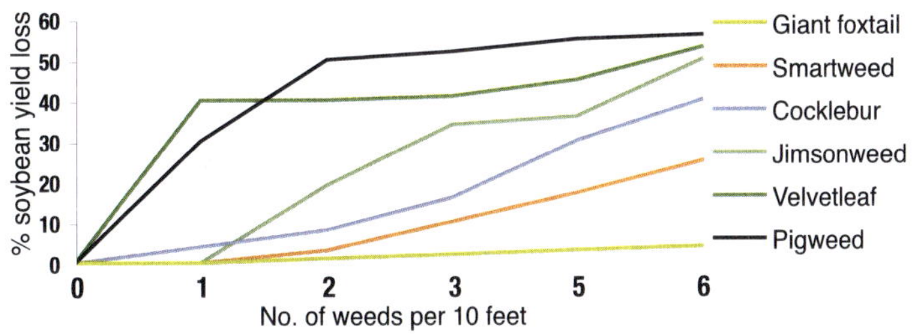

Approximate yield effects of early weed infestations in soybean. Giant foxtail reduces yield much less at low population levels when compared to pigweed or velvetleaf.

Weeds and crops are in constant competition in the field. Weed management is confounded by emergence, density, and diversity of species, but crops do have some innate tools against weeds. Crop density (planting rates, row spacing), competitive crops like rye and alfalfa, crop varieties de veloped for rapid canopy closure, rapid emergence, higher seedling growth rate, and weed tolerance are examples.

Changes in timing of tillage, planting date (early or delayed), increased crop rotation, increased crop variety, interseeding, etc. can break a weed cycle and lower the farmer risk of crop loss. Factors producers can manipulate include all of the following: planting date, cultivation, mulch, use of allelopathic crops, row spacing, planting density, intercropping, and selection of fastgrowing cultivars.

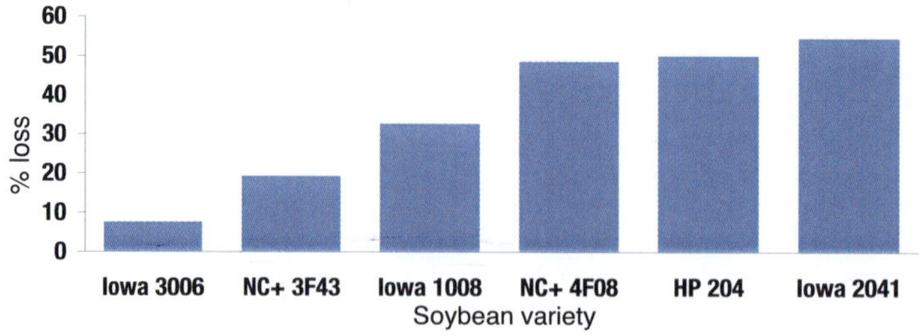

Yield loss due to weeds (as compared to weed-free controls) among six soybean varieties. Some varieties may yield better than others when in competition with weeds.

Reducing risk**:** crop competitiveness. Choose varieties and cultural practices that promote crop competitiveness.

Weed Identification

Dandelion

Type: Broadleaf perennial.

Where it grows: Lawns and gardens in sun or shade.

Appearance: This common lawn weed has a strong taproot; leaves are deeply notched. Yellow flowers mature to puffballs. Dandelion seeds are like parachutes that fly away in the wind—they're the plants that you would blow on and "make a wish" when you were younger.

Lawn Weed Control Tip: Mulch to prevent dandelions in gardens. Pull dandelion weeds by hand or use a postemergence herbicide in lawns.

Oxalis

Type: Broadleaf perennial.

Where it grows: Sunny or shady landscape, lawn, or garden areas.

Appearance: This garden weed has light green leaves that look like clover and cup-shape yellow flowers in summer and fall.

Control: Mulch garden areas in spring to prevent weeds. Pull oxalis weeds by hand or spray weeds with a postemergence herbicide in spring or fall.

Test Garden Tip: The leaves of oxalis are edible in small quantities and have a sharp, sour taste. However, plants should not be eaten if they have been treated with pesticide. The plant can be harmful if eaten in large amounts.

Crabgrass

Type: Grassy annual.

Where it grows: Lawn, landscape, and garden areas in sun or shade.

Appearance: Crabgrass is exactly what it sounds like: a grassy weed. This lawn weed grows roots anywhere the stem makes soil contact. Seed heads spread out like four fingers.

Control: Mulch your lawn to prevent crabgrass or use a preemergence herbicide; pull plants by hand or spot-treat with a nonselective postemergence herbicide.

Test Garden Tip: Each plant can produce up to 150,000 seeds.

Bindweed

Type: Broadleaf perennial.

Where it grows: Landscape and garden areas in sun.

Idnetify Weeds: Bindweed has arrowhead-shape leaves on twining vines. The garden weed bears white to pale pink morning glory-type flowers.

Control: Mulch your garden to prevent bindweed. Repeatedly chop down growing bindweed plants and/or treat with postemergence herbicide.

Test Garden Tip: Wandering roots produce offspring 20-30 feet from the mother vine.

White Clover

Type: Broadleaf perennial.

Where it grows: Lawn, landscape, and garden areas in sun to partial shade.

Appearance: White clover has three-lobe leaves frame round white flower clusters.

Control: Mulch your garden beds to prevent white clover in landscape areas. Use a postemergence herbicide in lawns or hand-pull the weeds.

Test Garden Tip: Clover adds nitrogen to the soil so as far as weeds go, this one is moderately helpful.

Nutsedge

Type: Grassy perennial.

Where it grows: Lawn, landscape, or garden areas with moist soil in sun or shade.

Appearance: Nutsedge has grassy leaves and nutlike tubers on the root system.

Control: Mulch garden areas in spring to prevent nutsedge. Pull plants by hand or spray with a postemergence herbicide containing monosodium methanearsonate (also known as MSMA).

Creeping Charlie

Type: Broadleaf perennial.

Where it grows: Shady lawn, landscape, or garden areas.

Appearance: Creeping Charlie groundcover has scalloped leaves and clusters of purple flowers in late spring.

Control: Mulch garden areas in spring to prevent creeping charlie. Pull plants by hand or spray with a postemergence herbicide in spring or fall.

Lamb's-Quarter

Type: Broadleaf annual.

Where it grows: Landscape and garden areas in sun or shade.

Appearance: Lamb's-quarter's scalloped leaves have gray undersides to them.

Control: Mulch your garden to prevent lamb's-quarter. Pull weed plants by hand or use a postemergence herbicide.

Plantain

Type: Broadleaf perennial.

Where it grows: Moist lawn and garden areas in sun or shade.

Appearance: Plantains have a broad, flat leaves around a low rosette.

Control: Mulch to prevent plantains growing in the garden. Pull these weeds by hand or use a postemergence herbicide in lawns.

Test Garden Tip: Each plantain plant can produce more than 15,000 seeds.

Dayflower

Type: Annual grass relative.

Where it grows: Sunny or shady landscape areas.

Appearance: Dayflowers have dark green leaves clasping a stem and brilliant blue flowers through the summer.

Control: Mulch the garden to prevent weeds or use a preemergence herbicide in spring. Pull weeds by hand or spot-treat with a nonselective postemergence herbicide.

Purslane

Type: Broadleaf annual.

Where it grows: Dry, sunny landscape and garden areas.

Appearance: Purslane groundcover has fleshy, dark green leaves and small yellow flowers at the ends of the stems.

Control: Mulch your garden to prevent purslane or use a preemergence herbicide in the spring. Pull plants by hand or spot-treat with a nonselective postemergence herbicide.

Velvetleaf

Type: Broadleaf annual.

Where it grows: Fertile, sunny landscape and garden areas.

Appearance: Velvetleaf gets its name after its large, velvety heart-shape leaves up to 10 inches across. The weed blooms yellow flowers in summer.

Weed Control: Mulch your garden to prevent velvetleaf or use a preemergence herbicide in spring; pull plants by hand or use a postemergence herbicide.

Wild Violet

Type: Broadleaf perennial.

Size: 6 inches tall, 6 inches wide.

Where it grows: Shady lawn, landscape, or garden areas.

Appearance: Wild violet is a groundcover with heart-shape leaves and purple flowers in late spring.

Control: Mulch garden beds in spring to prevent wild violet. Pull weeds by hand or spray with a postemergence herbicide in spring or fall.

Test Garden Tip: This plant is often grown as an ornamental in shade gardens.

Smartweed

Type: Broadleaf annual.

Where it grows: Sunny landscape and garden areas.

Appearance: Smartweed is an upright plant with pink or white flowers in summer and fall and has lance-shape leaves often marked with purple chevrons.

Control: To prevent this weed, mulch garden beds in spring. Pull plants by hand or apply a postemergence herbicide once it grows.

Test Garden Tip: This weed is native to areas of North America. Unlike many exotic weeds, it does support wildlife.

Quickweed

Type: Broadleaf annual.

Where it grows: Sunny landscape and garden areas.

Appearance: Quickweed has jagged, hairy leaves and small white daisy-shape flowers in summer.

Control: Use a mulch or a preemergence herbicide in spring to prevent quickweed. If plants do grow, pull them by hand or spot-treat them with a postemergence herbicide.

Pigweed

Type: Broadleaf annual.

Where it grows: Sunny landscape or garden areas.

Appearance: Pigweeds are tall plants with a taproot. These weeds sprout hairy-looking clusters of green flowers.

Control: Mulch garden areas in spring to prevent pigweed or use a preemergence herbicide in spring. Pull weeds by hand or spray with a postemergence herbicide.

Canada Thistle

Type: Broadleaf perennial.

Where it grows: Sunny lawn, landscape, or garden areas.

Appearance: Canada thistle has spiny, gray-green leaves and purple flowers.

Control: Mulch your garden to prevent it in landscape areas. Use a postemergence herbicide in lawns in spring or fall, or dig the weed out by hand.

Test Garden Tip: Thistle has an extensive root system that can grow several feet out from the main plant.

Knotweed

Type: Broadleaf annual.

Where it grows: Sunny or partly shaded lawn, landscape, or garden areas.

Appearance: Knotweed is an invasive groundcover with blue-green leaves sparsely appearing on long stems.

Control: Prevent knotweed with a deep layer of mulch or apply a preemergence herbicide in spring. Once the plant grows, hand-pull or spot-treat it with an herbicide.

Pokeweed

Type: Broadleaf perennial.

Where it grows: Sunny landscape or garden areas.

Appearance: Pokeweed bears light green leaves with clusters of white flowers and dark purple berries.

Control: Prevent pokeweed with a deep layer of mulch. Once the plant grows, hand-pull or spot-treat it with an herbicide.

Poison Ivy

Type: Broadleaf perennial.

Where it grows: Sunny or shady landscape or garden areas.

Appearance: Poison ivy can be a vine, shrub, or groundcover. The weed has leaves divided into three leaflets and can sprout clusters of green berries.

Control: Prevent poison ivy with a deep layer of mulch. If the weed starts to grow in your yard, spot-treat it with an herbicide.

Test Garden Tip: The plant contains oils that cause a severe allergic skin reaction in many people when touched. These oils are present even on dead leaves and can become airborne and inhaled if the plant is burned.

Black Nightshade

Type: Broadleaf annual.

Where it grows: Landscape or garden areas with rich soil in sun or shade.

Appearance: Black nightshade can be a bushy or climbing plant with white or purple flowers and purple or red fruits.

Control: Mulch your garden to prevent black nightshade. Pull the weed by hand or treat with a postemergence herbicide.

Test Garden Tip: All parts of this plant, including the fruits, are poisonous.

Black Medic

Type: Broadleaf annual or short-lived perennial.

Where it grows: Poor, dry, soil in full sun.

Appearance: Black medic has clover-type leaves and small, yellow flowers.

Control: Mulch to prevent black medic in gardens. Pull weeds by hand or use a postemergence herbicide. Discourage it by keeping the soil well watered and amended with organic matter (such as compost).

Quackgrass

Type: Grassy perennial.

Where it grows: Landscape and garden areas in sun or shade.

Appearance: Quackgrass has wheatlike flower spikes, which appear above slender clumps of grassy foliage.

Control: Mulch your garden prevent quackgrass. Dig plants out by hand, being sure to remove every bit of root.

Dock

Type: Broadleaf perennial.

Where it grows: Landscape and garden areas in sun or shade.

Appearance: Dock produces large, wavy-edge leaves and large seed heads covered with brown seeds.

Control: Mulch to prevent dock. Pull and dig up plants or treat with a postemergence herbicide.

Test Garden Tip: Each plant can produce up to 40,000 seeds that wait decades before sprouting.

Henbit

Type: Broadleaf annual.

Where it grows: Lawn, landscape, and garden areas in sun or shade.

Appearance: Henbit is a low, creeping plant with scallop-edge leaves and purple flowers.

Control: Mulch to prevent henbit in gardens or use preemergence herbicide in spring. Pull plants by hand or treat in lawns with a broadleaf, postemergence herbicide.

Fleabane

Type: Broadleaf annual.

Where it grows: Landscape and garden areas in sun to partial shade.

Appearance: Fleabane has slender leaves, borne on an upright stem that branches. It produces puffy white to pale lavender daisies.

Control: Mulch your garden to prevent fleabane or use a preemergence herbicide in spring. Pull plants by hand or spot-treat with a postemergence herbicide.

Nettle

Type: Broadleaf perennial.

Where it grows: Garden areas with rich, moist soil.

Appearance: Nettle has sawtooth-edge leaves and yellowish flower clusters covered with stinging hairs.

Control: Mulch to prevent nettle. Dig out weeds or treat with a postemergence herbicide.

Test Garden Tip: Always wear gloves when working around this plant.

Prostrate Spurge

Type: Broadleaf annual.

Where it grows: Lawn, landscape, and garden areas with dry soil.

Appearance: Green or purple-blushed leaves of prostrate spurge form dense mats.

Control: Mulch your garden to prevent prostrate spurge or use a preemergence herbicide in lawns. Pull weeds when young or spot-treat with a postemergence herbicide.

Chickweed

Type: Broadleaf annual.

Where it grows: Lawn, garden, and landscape areas with rich, moist soil in sun or shade.

Appearance: Chickweed creates lush green mats studded with small, star-shape flowers.

Control: Mulch to prevent chickweed in gardens or use a preemergence herbicide in early spring. Pull weeds by hand.

Test Garden Tip: Each plant can produce more than 15,000 seeds.

Musk Thistle

Type: Broadleaf biennial.

Where it grows: Landscape and garden areas in full sun.

Appearance: Musk thistle's prickly leaves off of tall stems topped by heavy 2-inch purple flowers.

Control: Mulch your garden to prevent musk thistle. Use a postemergence herbicide or dig the weed out yourself.

Ragweed

Type: Broadleaf annual.

Where it grows: Landscape and garden areas in sun or partial shade.

Appearance: Ragweed grows finely cut green leaves that are almost ferny.

Control: Mulch your garden to prevent ragweed. Use a postemergence herbicide or pull it out by hand.

Yellow Sweet Clover

Type: Broadleaf annual.

Where it grows: Landscape and garden areas.

Appearance: Yellow sweet clover has lanky branches, clover-like leaves, and fragrant yellow flowers.

Control: Mulch your garden to prevent yellow sweet clover. Pull plants by hand or spot-treat a postemergence herbicide.

Yellow Salsify

Type: Broadleaf biennial or short-lived perennial.

Where it grows: Sunny landscape and garden areas.

Appearance: Yellow salsify has gray-green leaves. Yellow flowers on the plant are followed by large puffballs.

Control: Mulch your garden to prevent yellow salsify. Pull plants by hand or treat with a postemergence herbicide.

Impact of Weeds

Weeds are one of the major threats to the natural environment. They are destroying native habitats, threatening native plants and animals and choking our natural systems including rivers and forests.

Directly or indirectly, all Australians are affected by weeds. Weeds reduce farm and forest productivity, invade crops, smother pastures and some can harm livestock. Land and water managers incur material and labour costs to control weeds - these costs are passed on to the Australian public through higher prices for produce.

The Economic Cost of Weeds

Weeds reduce the quantity and quality of Australia's agricultural, horticultural and forestry products, affecting both industry and consumers. It is estimated that weeds cost Australian farmers around $1.5 billion a year in weed control activities and a further $2.5 billion a year in lost agricultural production. The real cost of weeds to the environment is difficult to calculate, however it is expected that the cost would be similar to, if not greater than, that estimated for agricultural industries.

The Impact of Weeds on the Environment

Weeds are one of the major threats to Australia's natural environment. Major weed invasions change the natural diversity and balance of ecological communities. These changes threaten the survival of many plants and animals because the weeds compete with native plants for space, nutrients and sunlight.

Almost all of Australia's native vegetation communities have been invaded, or are vulnerable to invasion by exotic species that could result in changes to the structure, species composition, fire frequency and abundance of native communities.

Nationally, invasive plants continue to invade the land with exotic species accounting for about 15% of flora. About one-quarter of them are either serious environmental weeds or have the potential to be serious weeds.

Weeds may out-compete native plants because:

- They may not be affected by the pests or diseases that would normally control them in their natural habitats.
- The disturbed environment provides different conditions that better suit the invading weed.

As a result the weed may:

- Grow faster than native plants and successfully compete for available nutrients, water, space and sunlight.
- Reduce natural diversity by smothering native plants or preventing them from growing back after clearing, fire or other disturbance.
- Replace the native plants that animals use for shelter, food and nesting.

Weeds are often excellent at surviving and reproducing in disturbed environments and are often the first species to colonise and dominate in these conditions.

Weeds in the ocean spread over wide areas in a very short time. Introduced seaweeds came to Australia in the early 1980s in the ballast water of ships. They are now invading marine environments along the coast of south-eastern Australia.

The Impact of Weeds on Agriculture

Weeds reduce farm and forest productivity, they invade crops, smother pastures and in some cases can harm livestock. They aggressively compete for water, nutrients and sunlight, resulting in reduced crop yield and poor crop quality. For example, prickle bushes such as Gorse, Blackberries, Prickly Acacia, Parkinsonia and Mesquite can invade vast areas of grazing land preventing productive use of that land.

Weeds contaminate produce, for example:

- Burrs in wool contaminate fleeces.
- Grain milled with Saffron Thistle or Amsinckia results in discoloured flour.
- Animals that eat specific weeds, such as wild garlic, produce tainted milk and meat.
- Spines on fruit of Caltrop and Spiny Emex can damage the feet of stock animals.
- Paterson's Curse irritates the udders of dairy cows and can kill horses.
- Hemlock can be lethal to both stock and people.

Weeds can also affect the operation of farm machinery.

Farmers spend a large amount of time and money managing weeds. Despite control efforts, a recent Australian Bureau of Statistics survey of issues facing the agriculture industry found that weeds were the most commonly reported natural resource management issue affecting landowners. Weed-related issues affected 73% of Australian agricultural establishments during 2004-05. This compares to 46% of farmers reporting soil and land issues and 38% reporting water issues.

The Impact of Weeds on Human Health

Weeds can also cause human health problems. Many common weeds such as Parthenium Weed, Ragweed, Rye Grass and Privet cause asthma and other respiratory problems, especially in children. Some weeds can also cause skin irritation and some are poisonous.

Some water weeds such as Water Hyacinth (Eichhornia crassipes) and Cabomba (Cabomba caroliniana) can affect the quality of our drinking water if infestations are not managed within water supply dams.

There are no common characteristics of a poison or harmful weed that would help distinguish them. But as a general rule, plants with a bitter taste, unusual smell, milky sap or red berries may be poisonous with some plants having poisonous roots and bulbs.

References

- Plant: biologydictionary.net, Retrieved 2 March, 2019
- Plant-parts-functions: pmfias.com, Retrieved 7 January, 2019
- Different-types-plants-world, biology, science: zmescience.com, Retrieved 11 August, 2019
- Plantstructure, farm: waikato.ac.nz, Retrieved 18 February, 2019
- Plant-ecology: intechopen.com, Retrieved 20 July, 2019
- Weed, plant: britannica.com, Retrieved 25 April, 2019
- Introduction-to-weeds-what-are-weeds-and-why-do-we-care: extension.psu.edu, Retrieved 5 January, 2019
- Weed, plant: britannica.com, Retrieved 3 April, 2019
- Common-garden-weeds: almanac.com, Retrieved 22 March, 2019
- Plants-weeds-others-arent: todayifoundout.com, Retrieved 28 July, 2019
- Weed-biology: organicriskmanagement.umn.edu, Retrieved 8 May, 2019
- Weed-biology: organicriskmanagement.umn.edu, Retrieved 19 February, 2019
- Types-of-weeds, insects-diseases-weeds, pests, gardening: bhg.com, Retrieved 1 June, 2019

Chapter 2

Weed Control

The botanical aspect of pest control is known as weed control. Its aim is to stop weeds from contesting with desired flora and fauna. Weed control plays a vital role in agriculture and includes various types of methods like physical, chemical, cultural and biological weed control. This chapter discusses in detail the theories and methodologies related to weed control.

Weed control is vital to agriculture, because weeds decrease yields, increase production costs, interfere with harvest, and lower product quality. Weeds also impede irrigation water-flow, interfere with pesticide application, and harbour disease organisms.

Early methods of weed control included mowing, flooding, cultivating, smothering, burning, and crop rotation. Though these methods are still important, other means are perhaps more typical today, particularly the use of herbicide (plant-killing) chemicals. Another technique is to introduce insects that attack only the unwanted plant and destroy it while leaving the crop plants unharmed.

The inadequacy of the cultural, mechanical, and biological control systems, however, stimulated the rapid development of chemical usage since World War II. Herbicides have had an impact on crop production, changing many cultural and mechanical agricultural operations.

Herbicides are formulated as wettable powders, granular materials, emulsions, and solutions. Any of them may be applied as a spot treatment, broadcast, placed in bands, or put directly on a specific plant part. When formulated as solutions or emulsions, the chemical is mixed with water or oil.

Spraying is the most common method, permitting extremely small amounts to be applied uniformly because of dilution. Sprays can be accurately directed underneath growing plants, and calibration and rate control are easier with spray machines than with granular applicators. Granular formulations have advantages under some conditions, however. The use of herbicides must be integrated into the overall farm program because the optimum date and application rate depend on the crop stage, the weed stage, weather conditions, and other factors.

Careful use of herbicides in farm production lowers cost, resulting in a more economical product for the consumer. Herbicides cut the costs of raising cotton, for example, by reducing labour requirements for weed control up to 60 percent. Herbicides replace hand labour in growing crops, labour that is no longer available in developed nations at costs the farmer can afford. Machines for chemical application are widely available.

When used as directed, herbicides are generally safe, not only for the operator but also for wildlife and livestock. The greatest difficulty lies in accidental injury to crop plants resulting from drift and from residues in the soil, particularly if residues enter water courses.

The future of chemical pesticides and herbicides is under debate by those who manufacture, sell, and use them and by those who are concerned about environmental quality. The value of an assured food and fibre supply at reasonable cost is undeniable, and chemicals contribute much toward this. These substances also cause undesirable effects upon the environment, however, and indeed can be toxic to a wide range of organisms. This fact will demand an increasing amount of care in using chemicals, perhaps enforced by law, along with increasing use of nonchemical control techniques.

Controlling weeds in forage crop production may involve a wide range of techniques. Nevertheless, virtually all weed control methods may be classified into one or more of five main categories. The 5 general categories of weed control are:

Preventative Weed Control

Preventative weed control refers to any control method that aims to prevent weeds from being established in a cultivated crop, a pasture, or a greenhouse. Examples of preventative weed control would be using certified weed free seed, only transporting hay that is weed free, making sure farm equipment is cleaned before moving from one location to another, and screening irrigation water to prevent weed seeds from traveling along irrigation ditches.

Cultural

Cultural weed control refers to any technique that involves maintaining field conditions such that weeds are less likely to become established and/or increase in number. Examples of cultural weed control would be crop rotation, avoiding overgrazing of pastures or rangeland, using well-adapted competitive forage species, and maintaining good soil fertility.

Mechanical

Mechanical weed control refers to any technique that involves the use of farm equipment to control weeds. The two mechanical control techniques most often used are tillage and mowing.

Biological

Biological weed control refers to any technique that involves the use of natural enemies of weed plants to control the germination of weed seeds or the spread of established plants. This is a rapidly expanding area of weed control with many examples. Examples of biological weed control include sheep to control tansy ragwort or leafy spurge, cinnabar moth and the tansy flea beetle to control tansy ragwort, the chrysolira beetle to control St. John's Wort, and the use of goats to control brush on rangeland.

Chemical

Chemical weed control refers to any technique that involves the application of a chemical (herbicide) to weeds or soil to control the germination or growth of the weed species. In economic terms, chemical control of weeds is a very large industry and there are scores of examples of chemical weed control products. Common examples of chemicals used to control weeds in forages are 2,4-DB; EPTC; bromoxynil; and paraquat.

Weed Management

Although weeds are present in every orchard, there are wide variations in the species growing and density of each population. Just as scouting for insects and diseases is well established in integrated pest management, scouting for weeds is the basis for Integrated Weed Management (IWM). Information gathered from weed scouting allows growers to:

- Identify weeds present early in the season when they can cause yield losses.
- Match herbicides to weeds.
- Choose cultivator for weed stage.
- Alter cultural practices for different weed life cycles.
- Discover weed patches before they spread through field.
- Identify areas to avoid cultivation.
- Identify areas for spot treatments.
- Choose the optimum timing for maximum control.

How to Scout for Weeds

Scouting for weeds can be done while looking for insects and diseases, although a separate walk through the orchard allows for more detailed observations and collections. Basic first steps are:

- Identify type of weed - e.g. grass or broadleaf.
- Identify life cycle - e.g. annual, biennial, perennial.
- Identify name of weed, if possible.
- Identify source of weed - e.g. mulch, fence row, weed escapes, irrigation water.
- Take samples for verification with roots, flowers and shoots - in paper bag inside a plastic bag, with a label.
- Work a zig-zag pattern through the field - remember weeds grow in patches.
- Check field edges for weed invaders.

How to Record Weed Scouting

Recording information on scouting reports helps make decisions this season, and provides a long-term record of weed emergence patterns and problems in the orchard. Record the following:

- Identify the weed, if possible.
- The growth stage of the weed and crop.
- The number and shape of the leaves.

- The stage of flowering.
- Areas and percentage of the field infested - mark on a field map.

Tools Needed for Weed Scouting

To do a good job of field scouting, bring along:

- A keen eye
- A hand lens
- Field maps
- Weed identification books
- Bags for samples and marker
- Field history

Basic Biology of Weeds for Field Scouting

It is easy to distinguish between broadleaf and grassy weeds. Beyond this, learn the growth habit of the weed, and target management strategies at susceptible stages.

Weeds have one (or more) of the following life cycles:

- Annual weeds - grow and flower in one year. Some weeds in orchards are winter annuals, i.e. they begin their growth in the fall, forming a rosette, and flower the following spring or summer.
- Biennial weeds - have a two-year life cycle, producing leaves in the first year and flowering in the second year.
- Annual and biennial weeds compete for nutrients and water as they grow under trees. After they flower, they die. However, their seeds may cause recurring problems for several years by forming a soil seed bank.
- Perennial weeds - live for many years, and generally establish from various types of root systems, and many also spread by seeds. They usually flower every year, expanding their root system and spreading by both methods through orchards. Perennial weeds can be very competitive, especially if they grow in thick patches.

Table: Common types of weeds in orchards.

Annual	Biennial	Perennial
Barnyard grass	Burdock	Canada thistle
Common chickweed	Buttercup	Creeping Charlie
Crab grass	Pepper grass	Dandelion
Downy brome (grass)	Wild carrot	Milkweed
Lamb's-quarters	Yellow rocket	Plantain

Pigweed		Poison ivy
Ragweed		Quack grass
Smartweed		Wild grape

Goals of Weed Management

The goal in managing weeds in tree fruit orchards is to maximize yields by suppressing weed competition during critical periods of crop development. This concept is the critical weed-free period - when it is most important to control weeds to prevent competition with trees. It is important to understand these facts about the critical weed-free period:

- If the crop is kept weed-free for the critical weed-free period, no yield reduction occurs.
- Weeds emerging after the critical weed-free period do not affect yield.
- When weeds compete with trees during the critical weed-free period (especially during the planting year), effects may last several years, and possibly for the life of the orchard.

There are other valid reasons to control weeds outside of these critical periods such as attractiveness for pick-your-own customers, harvest efficiency and reduction of weed seed banks. However, controlling weeds outside of the critical weed-free period does not increase yields.

Weed Competition Effects

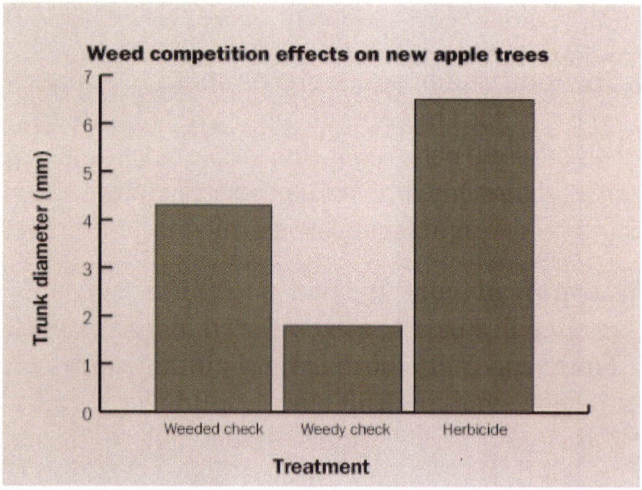

Weeds competing with newly planted apple trees cause severe growth reduction in only three months.

Weed competition in the planting year drastically reduces tree growth. After tree planting, it is challenging to maintain a weed-free strip under trees, but even a small amount of weed growth reduces tree growth. shows the reduction in tree growth caused by weed competition during the first three months (from planting in May until July). The herbicide treatment maintained a complete weed-free strip from May until July, and resulted in the most tree growth. Trees in the hand weeded check had less growth, because of some weed growth between hoeings, emphasizing the need to maintain a weed-free area. Other tree research shows this effect remains with the tree for many years.

Critical Weed-free Period for New Orchards

The critical weed-free period for new orchards is the first three months after planting. Research on weed effects on newly planted apple trees shows the need to control weeds during May, June and July. shows that controlling weeds in May allows for the most tree growth compared to controlling weeds during June only or July only. If the length of the weed-free period is extended to May and June, tree growth is greater than during only one month, and is also greater compared to June and July, or July and August. Maximum growth is achieved when the under tree area is kept weed-free for the three months of May, June and July.

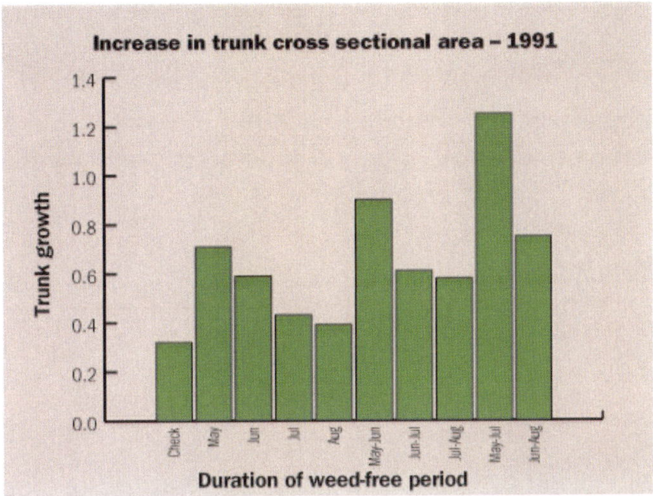

Tree growth in newly planted Gala/M26 trees under one month, two month or three month weed-free periods.

The figure below shows results from the same trees when they produced their first harvest two years later. Note that trees kept weed-free for May only, May and June, or May, June and July produced the most yield per tree in their first harvest. This indicates that the critical weed-free period for newly planted trees is the first three months after planting.

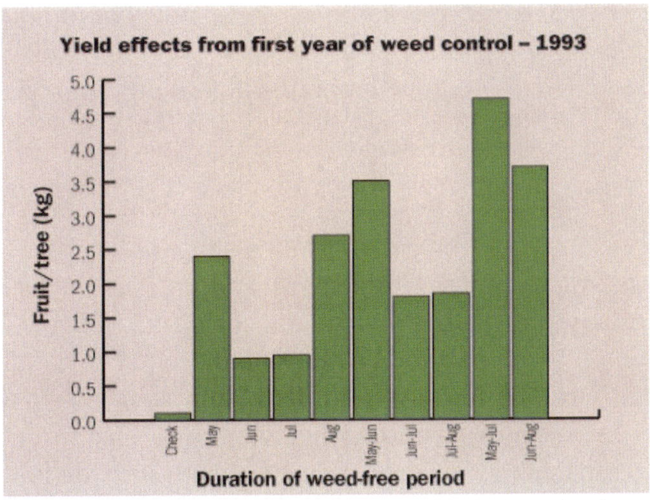

The yield on young Gala/M26 trees is directly related to the length of weed-free period in the planting year - the greatest yield was produced when trees were kept weed-free during May, June and July of the planting year.

Critical Weed-free Period for Bearing Trees

For bearing apple trees, the effects of weed competition are the greatest from bud break until terminal bud set, i.e. spring until early July. This corresponds with four stages of crop development:

- Flowering
- Fruit set
- Fruit enlargement
- Flower bud initiation

Conditions during flowering, fruit set and fruit enlargement effect this year's crop, and conditions during flower bud initiation affect next year's crop. Weed competition must be suppressed during this critical weed-free period - bud break to July. Inadequate grass seeding between the rows may result in early weed problems in young orchards.

Weed competition early in the growing season robs trees of ground-applied nutrients - note the chickweed and shepherd's purse growing under the trees during the critical weed-free period.

Improper or inadequate grass seeding between tree rows results in early weed problems in a young orchard.

Benefits of Weed Competition

Trees prepare for winter's cold by hardening off, starting in the late summer and through the fall. At this time, some small weed growth under the trees is acceptable, and may encourage trees to harden off. Weedy cover later in the season also reduces the risk of soil erosion and creates overwintering habitat for beneficial insects.

A well-established orchard sod with minimal weed growth - allowing weed growth in herbicide strips late in the season may help trees "shut down" for winter.

Remember that weed suppression does not mean weed-free. Suppression of weeds to prevent competition for moisture and nutrients is usually enough to maximize crop yields. Where irrigation is used and nutrient levels are high, trees tolerate a higher level of weed competition than low nutrient or drier soils.

Early weed escapes in the planting year compete with young trees for moisture and nutrients - the addition of water through irrigation (note irrigation tubing) allows trees to tolerate more weed competition.

Weed Management Tools

Herbicides are one of the tools available to control weeds in orchards. Mulching, mowing and cultivation are also effective in managing weeds, especially in the year of planting. Each strategy has certain strengths and weaknesses as indicated in table below.

Table: Advantages and disadvantages of weed management tools.

Tool	Advantages	Disadvantages
Cultivation/ hoeing	- Effective, especially on small weeds - Non-selective - controls all emerged growth - Equipment readily available	- May damage soil structure - May spread perennial weeds - May damage trees/roots - Provides only short-term control

Mulching	- Effective if properly managed - Non-selective - suppresses all emerging weeds - Holds soil moisture as well - Provides long-term control	- Availability of mulch - Cost of mulch/application - Attractive to rodents - May affect tree nutrition - Must be free of weed seeds
Mowing	- Rescue treatment - Quick suppression - Equipment available - Reduce seed spread	- Weeds may still compete - Quick regrowth - Several mowings required - May damage young trees
Herbicides	- Effective - Easy to apply - Can be selective - Timely	- Require 2% soil organic matter - Directed spray equipment - Effects on pest complex - Cost varies

Weed management strategies may combine several of these tools and it is important to be aware of the pros and cons of each. Mulching, cultivation and/or mowing may affect the performance of herbicides applied under trees, as well as using irrigation.

Weed "escapes" can occur around trickle irrigation emitters due to leaching of herbicides.
Use spot treatments if escapes occur during the critical weed-free periods.

How Herbicides Work

Herbicides registered for orchards kill weeds in several different ways, but there are three broad categories of control methods:

- Burndown (contact) - herbicide is applied to existing vegetation and kills the top growth. Herbicides are directed under the trees to create a weed-free strip and/or applied to the row alleyways. Some burndown herbicides are systemic - they are absorbed into the plant and move to the growing point and/or the roots, and give longer-term control for perennial weeds. Some burndown herbicides are not systemic and target only above-ground green tissue. These contact herbicides control top growth, but do not give long-term control of perennials.

- Soil residual (pre-emergent) - herbicide is applied to the soil surface before weed seedlings germinate. The chemical remains in the soil for several weeks to months, killing germinating seedlings. Generally, a rainfall of at least 12 mm is required to activate the herbicide, resulting in a longer period of control.

- Post-emergent (selective) - these herbicides kill only certain weeds out of existing vegetation e.g. graminicides kill grasses only, or auxinic (hormone) herbicides kill only broadleaves. These are usually targeted for weed escapes or for problem weeds like thistles or annual grasses.

There are advantages and disadvantages to using these types of herbicides as indicated in table below.

Table: Advantages and disadvantages of herbicide control categories:

Category	Advantages	Disadvantages
Burndown	- Controls all emerged weeds - Inexpensive - Widely available - Can target critical weed-free period	- May damage trees if absorbed (systemic) - Early weeds compete with trees - Perennial weeds are not killed - New seedlings germinate after application
Soil residual	- Reliable control - Effective for longer periods - Can target critical weed-free period - Cost effective - Broad spectrum - Longer window to apply	- May cause tree injury - Not safe on low organic matter soil - May have tree-age restriction - May leave residues after orchard removal - May require incorporation - Applied before weed problems are known - May affect soil biology
Selective	- Can target critical weed-free period - Targets specific weeds - Minimizes herbicide use - Generally safe for trees	- Does not control a broad spectrum of weeds - Often an extra application - Additional expense - Timing is critical

Herbicides are grouped by modes of action by the Weed Science Society of America (WSSA), and are listed in the OMAFRA Publication 75, Guide to Weed Control. For each of the herbicide categories listed, there may be herbicides from several mode of action groups. For example, glyphosate (e.g. Roundup, Touchdown, Vantage) and paraquat (Gramoxone) are common herbicides used for burndown, but glyphosate is a Group 9 herbicide and paraquat is a Group 22 herbicide.

Be sure to know both the control category and herbicide group for herbicides used in the orchard. For example, a weed management strategy may require a burndown herbicide in the spring may use a Group 22 herbicide because a Group 9 herbicide was used the previous fall.

Rotating herbicides with different modes of action is an important strategy to manage weed resistance.

Many herbicides only control grasses or broadleaf weeds. For broad-spectrum control of weeds usually growing in orchards, apply a tank mix of two herbicides. Choose products with strengths on different weeds, e.g. Dual II Magnum (controls grasses well) is often tank mixed with Princep (controls broadleaf weeds well). If weed growth is present, include a burndown herbicide in the tank mix.

Selective herbicides are useful for treatments on weed escapes, for specific problem weeds like grasses or vetch, or for reducing pressure from broadleaf weeds with fall treatments of 2,4-D. Waiting until weeds appear before using herbicides is a preferred approach in a truly integrated weed management, as long as practical considerations such as timeliness of application, resistance management and critical weed-free period have been considered.

Issues with using Herbicides

Managing Weed Resistance

There is concern that continuous use of the same herbicide selects for resistant weeds, and herbicide resistance has been identified in a growing number of weeds and locations around the world. In Ontario orchards, resistance to paraquat has been identified where repeated applications each year over several years was practiced.

In orchard practice, it is common to use herbicides with different modes of action, often during the same growing season. For example, triazine herbicides (eg. Princep/simazine) have been widely used in orchards, creating concern of selecting for triazine-resistant pigweed and lamb's-quarters. Corn growers had previously selected triazine-resistant weeds through continuous use of triazines alone. However, when orchard growers include other herbicide modes of action, e.g., Roundup, Ignite or Gramoxone, any resistant weeds are killed by these treatments. The escaped weeds are usually species that simazine does not control well e.g. pigweed and crabgrass. Only tests confirm if weeds are truly triazine-resistant. With more herbicides available to apple growers, choose different herbicide families and modes of action. Rotate herbicides within each year, and between years, in each orchard block.

In many orchards, pigweed is a challenging weed, which often escapes after simazine is applied.

Contamination of Surface and Ground Water

Many herbicides are water soluble, and may leach through the soil profile, eventually reaching the ground water. Heavy rainfall or irrigation immediately after application contributes to this problem. Herbicide spills near wells are another potential source of ground water contamination. Avoid these situations, especially on low organic matter soils.

Soil residual herbicides such as simazine bind tightly to soil particles and may be washed into surface water when soil erosion occurs. Practice erosion control if the orchard site is at risk of wind or water erosion.

Remember herbicides lost through leaching or erosion are not available to control weeds in your orchard.

Managing Spray Drift

As spray applicators, growers are required by the Pesticides Act to ensure no pesticide lands off the intended target. There are two types of spray drift to avoid:

- Physical drift: When tiny droplets (less than 100 microns) are moved by wind and air currents off-target.
- Vapour drift: When volatile products (e.G. 2.4-D) become a vapour and move off-target.

The Ontario Grower Pesticide Safety Course suggests 10 ways to reduce spray drift:

- Use anti-drift technology - nozzle hoods, boom hoods, perforated screen, air-assist/curtain and shields.
- Use alternative application technology, e.g. wick wiper.
- Where possible, don't spray - use cultivation, flamers, mulch biological control or mowers.
- Read the label and follow directions for stage of growth, weather and other precautions.
- Watch the weather - and avoid spraying in winds over 10 km/h, temperatures above 25°c and low rh (<75%).
- Use buffer strips, check label for required distance - if not listed, stay back at least one spray boom width.
- Use high water rates and larger nozzles - results in larger droplets and better spray coverage.
- Use adjuvants with caution - check label, may distort spray pattern.
- Lower the boom, may need wider angle nozzles.
- Choose the right nozzles - drift reducing styles are now available.

Mount shields to protect nozzles from wind and prevent drift - an added benefit is protecting low branches from herbicide.

Mount wick wipers to swipe weeds growing under trees without tree contact or herbicide drift.

Use mowers with wing arms to cut weeds around tree trunks - a sensor kicks out the mower arm around the tree to prevent mechanical injury.

Effect of Herbicides on Non-target Pests

Herbicides can affect organisms in orchards, not just weeds:

- Some herbicides are toxic to soil fauna like earthworms, e.g. Linuron (lorox).
- Some insect pests thrive on weeds, control these weeds to reduce fruit damage, such as:
    - Tarnished plant bugs thrive on creeping charlie and chickweed - control these weeds to reduce injury.
    - Twospotted spider mites overwinter on the ground cover and build up populations on these weeds in the spring - remove host weeds early in the season to reduce problems.
    - Caution - where tarnished plant bug or twospotted spider mites are established on the orchard ground cover, avoid killing the weeds with herbicide and forcing pests into trees.
- Controlling dandelions in orchard sod (e.g. with 2,4-D) may affect bee activity during pollination time - honeybees appear to prefer apple blossoms to dandelions but bees may use dandelions to maintain hive strength in spring if flying conditions are poor during apple bloom.

Crop Safety

Herbicides can cause injury to apple trees if used improperly. Read the label to determine restrictions on soil type or age of tree.

- Soil organic matter: Generally soils with at least 2% organic matter (OM) are required to use soil residual herbicides. For low organic matter soils (OM <2%), choose Devrinol because its crop safety is not dependent on OM. Mulch is also a good choice for low OM soils. A soil OM of 3% is recommended to avoid crop injury when using Sinbar, because it tends to be less bound to soil particles.

- Age of trees: Some herbicides are only registered for new trees. For Treflan, this restriction is because the product must be incorporated. For other products, there has not been research to prove there is no residue in the fruit. Some herbicides are only used on older trees. For example, linuron is restricted to some tree species more than 10 years old because the original research was only done on trees 10 years old. In some cases, there is a need to wait until the trees have a larger root system or thicker bark to avoid tree damage.

Herbicide Residues in Soils

For orchards scheduled for removal in the next three years, avoid soil-applied herbicides with a long carryover period. Do not use simazine (Princep, Simadex), Devrinol and Sinbar for several years before orchards are removed.

Soil Incorporation Requirement

Devrinol and Treflan require incorporation into the soil to prevent sunlight from breaking them down. For Treflan, cultivation with tillage equipment is required, restricting its use to newly planted trees. For Devrinol, incorporation by rainfall or irrigation must take place within two days in the summer or seven days in the late fall.

Cultural Weed Control

Organic farmers recognise that every element of farming is inter-linked, and that good rotational design produces healthy soil, healthy plants and good yields. Crop rotation is the cornerstone of organic farming practice. Rotation and forward planning are also important for managing weeds. In this section we provide information on cultural methods aimed at preventing weed problems arising in the first place and which farmers and growers can plan to incorporate into their rotations.

The underlying principle of a preventative approach is to produce a constantly changing environment to which no single weed species can adapt and become dominant and unmanageable. In practical terms this means as diverse and long a rotation as possible consistent with the farm system and which prevents the weeds returning seeds to the soil seed bank.

Crop planning is a cornerstone of organic farming practice and it has important implications for weed management. It can be designed to positively influence weed control and to make a useful contribution to the whole farm management strategy. Typically rotation cycles extend over several years with often only an annual change of crop, but the inclusion of cover crops, intercrops and green manures increases the crop diversity in a rotation. In horticultural systems there may be sequential cropping where short-term crops follow each other in succession.

Weed population density may be markedly reduced using crop rotation but there has been little experimentation. Success depends on the use of crop sequences that create a diverse pattern of competition, allelopathic interference, soil disturbance, and production needs (such as the time of sowing and harvesting). There should be regular changes between spring and autumn-sown crops, and between annual and perennial crops, between dense leafy crops and those with an open habit, and between crops that require a long growing season and others that mature quickly. Rotation may also allow the use of a range of cultivations and direct non-chemical weeding methods that may be applicable to the different crops. The aim is to provide an unstable and inhospitable environment that prevents the proliferation of a particular weed species.

Choosing Crops and their Sequence

The length of the rotation, the choice and sequence of crops will depend upon individual farming circumstances that will include factors like soil type, rainfall, topography and enterprises. However, the aim is to produce an unstable environment in which no single weed species is allowed to adapt, become dominant, and therefore difficult to manage. No one rotation can be recommended, but ideally in terms of weed control rotations should include:

- Alternation of autumn and spring germinating crops,
- Alternation of annual and perennial crops (including grass),
- Alternation of closed, dense crops such as oats which shade out weeds, and open crops such as maize which encourage weeds,
- A variety of cultivations and cutting or topping operations that directly affect the weeds.

Various suggestions and observations include:

- Putting sensitive annual crops after perennial leys. Research has shown that in the third cropping year after a grass/clover ley there is twice as much weed emergence as compared to the first.
- Include a row crop that allows the use of one or more cultivations to kill emerged weeds and encourages the germination of others, so reducing the soil seedbank and hence potential weed numbers in future crops. Cultivations may also reduce the problem of perennial weeds by disrupting growth and smothering regeneration in the growing crop. Typical cleaning crops include turnip, sugar beet, and potato.
- Uncultivated leys provide a completely different habitat for weeds and may be used to reduce or eliminate particular weed species. Few studies have been made of the effectiveness of leys for controlling weeds but trials suggest that there is little advantage for weed

management in leaving leys down longer than 3 years. The species composition, and the mowing and grazing regimes are important in the weed dynamics. Management of the weeds at the time of ley establishment is critical as is the method of ending the ley to avoid a flush of weeds due to the release of seed dormancy by cultivation. A greater proportion of ley in the rotation usually results in lower seed numbers in the seedbank in comparison with arable crops. It was a traditional way to deal with land infested with wild oat but does not eliminate the weed completely.

- Where a long grass break does not form part of the rotation weed problems are likely to be more severe. The problem will be greater where less vigorous and therefore less competitive crops are grown. Among the cereals, oats and winter rye are the most competitive followed by triticale.
- Canopy development and shading are important for weed suppression and choice of cultivar can influence this.
- Higher seeding rate and narrower row spacing increase the level of weed suppression.
- Competitive cereals like rye may be grown as short duration.

Fertility Building

The fertility-building period, or ley, will influence the weed population. If it is well managed it can act as a weed suppressing phase. It is important to choose the right species and ensure they establish quickly, especially for grassland systems where the ley may last for several years. Establishment of leys can be easier in the autumn period than in the spring because sowing in spring coincides with the main spring flush of weeds. The seedbed needs to be well prepared, and good contact made between seed mix and, ideally, moist soil to achieve good establishment. The choice of fertility building crop is also important. Rotations with grass leys have been shown to be beneficial in reducing weed seed numbers compared with rotations that do not include a grass phase. Grassland systems, which have temporary leys rather than permanent pasture, will provide the opportunity to control perennial weeds during the cultivations between ploughing and reseeding.

Grassland or clover/grass leys are an important part of the organic farming system in the UK. On livestock farms grassland forms the basis of the production process, in arable systems the ley is used primarily for maintaining or restoring soil fertility. The grass may be managed as a short, medium or long-term crop and this may determine the composition of the desirable sward species and the nature of the associated weeds. The seed mixture for a ley may include a relatively simple mixture of grasses and legumes or may be more complex and contain a range.

Choosing Varieties

Varieties that consistently suppress weeds are generally more desirable in organic systems (although this might be outweighed by marketing necessities) as opposed to varieties that tolerate weeds (and which potentially allow weeds to develop and return seeds to the soil seed bank).

- Organic varieties: Should show quick germination and establishment, rapid early and vigorous growth, and the ability to rapidly cover the soil and shade it (prostrate or tall varieties) to out-compete weeds at an early a stage in the crop cycle as possible. Varieties are well known to differ in architecture and competitive ability and whilst those that out-compete weeds are preferred it should also be borne in mind that those with erect foliage or that can tolerate some degree of mechanical weeding are also likely to be useful. Some crop types or varieties might produce allelochemicals that prevent weeds from developing or germinating although information on this is generally lacking.

In grassland systems the choice of variety may be dominated by forage value, but if there is opportunity the most vigorous species should be selected, as these will determine the productivity of the whole ley period. The trend in conventional cereal production has been to grow the taller stemmed varieties for their weed suppressing ability. Some farmers have stayed with the shorter stemmed varieties and employed a weed topper/cutter which will remove and, ideally, collect weed seedheads so long as there is a difference in height between crop and weed. New research in wheat is investigating leaf angle development, height and speed of development on weed suppression to aid farmer variety choice.

- Seed size: Varieties with a larger seed size have been shown to exhibit greater initial vigour of emergence and growth, which may subsequently provide extra competitive ability. If there is a choice available then the most vigorous species should be selected, as these will be more likely to out-compete weeds and suppress their development.

- Clean seed: It is important that the crop seed is free from contamination by weed seeds. Organic farmers are obliged to use organically produced seed and this should be clean. It is important if saving seed that it is taken from weed free crops, and ideally, professionally cleaned. Tolerance levels of contamination should be low although they are not generally well defined as of yet. Tolerance levels of dock in Switzerland are one seed per 100g.

Seed Rate and Crop Spacing

Spatial distribution of the canopy foliage and rooting system will be important for weed suppression. In drilled or transplanted crops the proximity of the plants to one another will determine the competitiveness of the plant stand as a whole. The principle is that the greater the amount of space taken up by the crop in the rows, the less space there is available for the weeds to invade. However, it should be borne in mind that closely spaced crop plants compete with each other and that it is also expedient to allow sufficient space between plants to allow for efficient mechanical weeding should weeds develop and threaten the crop.

Seed rates tend to be higher for organic than conventional crops. There is also an allowance for potentially lower germination rates and loss of the crop by mechanical weeders.

There has been much work in cereals on row spacing, pattern, direction of sowing and seed rates (typically 10% higher in organic cereals). Results are varied and interactions between the factors are often significant. For example in narrow widths an E-W sowing was favourable whereas in wide row widths a N-S sowing showed better response. Findings from the EU funded WECOF trials are awaited to give more definite recommendations.

In vegetable crop market size specifications are often the main driver in determining the crop row spacing.

Establishing the Crop

The ability of the crop to get off to a good start ahead of the weed flora is critical. Good soil management practises are important to provide the best possible seedbed in which to plant a crop. The impact of a poor compacted soil can soon be seen on crop establishment and subsequent weed invasion.

In some systems sowing can be aided by the use of primed seed, or by transplanting an already established plant into a freshly prepared weed free seedbed.

Transplanting is a popular technique in organic horticultural systems. Bare rooted transplants can be raised on holdings or modular plants raised or bought in then planted out in the field. Advantages include the benefit of accurate spacing i.e. not having to rely on germination that can sometimes lead to uneven establishment with subsequent yield and quality penalties. It also accentuates the difference in size between crop and weed, which can be vital for mechanical weeding at later stages.

Intercropping and Undersowing

Intercropping (or mixed cropping) and undersowing involves growing two or more different crop species in the same area. The advantage for weed control is that the crops cover more ground, so there is less space available for weed emergence. Intercropping can involve purely cash crops or a mixture of cash crops and fertility building crops.

- Intercropping: Is widely practised in certain countries and an enormous variety of intercropping systems have been developed. Both component crops may be taken to yield or one may be there as a living mulch to improve weed control. In successful intercrops weed suppression is usually superior to that of either of the component crops when grown alone. Crop density, crop diversity, crop spatial arrangement, choice of crop species and cultivar will all affect weed growth in intercropping systems. If water or nutrients are limiting then growth of one or both intercrops will suffer. Improved weed control alone is unlikely to justify their use and there must be other obvious benefits if the change in cropping practice is to prove economic.

In terms of mixed cash cropping there have been investigations into organic winter wheat and beans that reduced weed growth and gave better yields than sole cropping. Leeks and celery intercropping has also been shown to increase weed suppressing ability, and reduce reproductive capacity of late emerging groundsel. There are probably a wide range of combinations that could be designed to suit the farmer's rotations and marketing needs although some practical experimentation is required.

- Undersowing: Aims to cover the ground with a quick-growing dense layer of vegetation underneath the crop. The undersown species is prostrate, usually leguminous, and adds to or maintains fertility. It also suppresses weeds. Combining cash cropping with fertility building in this manner potentially produces an economic return, and it may mean there is either no need for an isolated fertility period, or that the length of that phase can be reduced. Undersowing cash crops with fertility building crops has other advantages apart from weed suppression, but so far the technique has only been widely used in cereal growing where it helps to re-establishing leys and avoid bare ground after harvest. Short-strawed cereal varieties can be difficult to manage, and straw difficult to save, due to the undersown crop growing up into the cereal. Further research into variety choice and sowing rates is needed to ensure that competition between the two crops is kept to a minimum.

Using Cutting Regimes for Weed Control

Cutting and topping weeds will have an impact on the type and weed flora in a field and can be invaluable in preventing return of weed seed to the soil seed bank. Cutting and topping are important for weed management in pasture, grass and leys. Topping can also be used as a remedial measure in vegetable or other crops to prevent weeds from seeding.

- Pasture systems: Good management involves maintaining the condition of the sward by cultural means. In particular reducing weed intrusions by chain harrowing in spring and topping regularly during the growing season. Mowing during the seeding year must be carefully judged and close cutting avoided. Spring sown stands should be cut no later than mid-August to allow recovery before winter. Summer sowings should be left unmown until November. Undersown lucerne should be left to grow into the winter. Where companion grasses are growing strongly, light winter grazing may be desirable. In grazed pasture weeds that are not eaten by livestock, will need to be topped to prevent seed shed. The established crop may be cut up to four times per year starting in mid-May. The crop is quickly weakened by defoliation, either by grazing or cutting, at too young a stage especially in spring or autumn. Before entering the dormant stage the crop must be allowed to make sufficient growth to replenish the food reserves in the root.

- Grassland systems: Cutting for hay or silage will have an impact on the weed flora. Silage tends to be cut early in the season when the sward is young and fresh, whilst hay is cut at a later stage. There can be both advantages and disadvantages associated with the timing of cutting depending on the weed flora and the ultimate requirements of the system. Cutting late may allow weeds in the pasture to grow to maturity and set seed. The ripe seeds may contaminate the hay and remain viable when passed through livestock. Dock seeds should not survive low pH silage, however they will survive in a later cut of hay. This mature seed may also shed on the ley surface and find opportunities to germinate in situ or be transported by livestock to other locations. In contrast, cutting early for silage in fields, with for example an infestation of creeping thistle, may encourage the spread and growth of this weed. Hence, there has to be a balance between the requirements of the farming system and weed control implications.

- Horticultural and stockless arable systems: Ley management will include topping at intervals during the summer to a height of around 10-15 cm. Ideally in fertility building leys the sward should not be allowed to get higher than 40 cm (or knee height). If the vegetation gets higher than this, then topping will create a mat of vegetation that will act like a mulch. This can create dead spots in the ley where clover may be excluded by the more vigorous grasses, or which weeds may colonise. Topping the ley regularly will also ensure that tall weeds that may have germinated will not be able set seed.

Using Manures

The use of raw manures and slurry has often been associated by farmers with increased weed problems. The problems can arise in various ways; either as a result of weed seeds in the manure, as a result of the way in which it is applied or due to the stimulatory effect of the nutritents on weeds already present in the soil.

- Weed seeds in manure: Some manure contains weed seed, either seed that has passed undigested through animals or from bedding materials like small-grain straw and old hay. High-temperature aerobic composting (recommended under organic standards) can greatly reduce the number of viable weed seeds as long as the temperature is maintained at higher than 60 °C for more than three days. Operationally compost will need to be regularly turned to achieve even heating through the whole heap and to get material from the outside (where seeds are likely to survive) to the inside (where the highest temperatures are likely to be generated. In a similar way, aeration of slurry can reduce the number and viability of weed seeds.
- Applying manure: When applying manure or slurry try not to create conditions which stimulate weed seeds to germinate (excessive soil disturbance, creating bare patches etc.). Applying slurry to stubble after silage cuts can provide optimum conditions for weed seed germination. A nutrient-rich bed of cattle slurry will produce a high potassium environment which will favour weeds such as docks rather than grasses. Dock seeds should not survive low ph silage but will survive in a later cut of hay.

Some research has shown that placing manures and slurry more accurately on crops can benefit the crop rather than the weeds. Crop plants are generally sown fairly deeply and they germinate from a lower level in the soil profile than weeds, which tend to dwell on the surface and germinate from 0-3 cm. Crop plants also root more deeply. This tendency can be exploited for weed management. In arable/horticultural systems manure placed 10 cm below the soil surface encourages the crop seeds to grow down into the nutrient-rich layer before the surface-dwelling weeds can reach it. This technique can also be used with broadcast spread slurry. If it is ploughed in rather than left on the surface it will be available to the crop before the weeds can reach it.

In many cases, the growth of weeds that follows manuring is a result of the stimulating effect manure has on weed seeds already present in the soil. This can be due to the flush of nutrients (e.g. supply of nitrates), enhanced biological activity in the soil, or other changes in the fertility status of the soil. Some work has indicated that excesses of potash and nitrogen in particular can encourage weeds but in any case it is prudent to monitor the nutrient content of your soil and manure, and spread manure evenly to reduce the incidence of weed problems.

Don't give the docks an advantage: applying slurry to stubble after silage cuts can provide optimum conditions for weed seed germination. A nutrient-rich bed of cattle slurry will produce a high potassium environment which will favour weeds such as docks rather than grasses. Dock seeds should not survive low ph silage but will survive in a later cut of hay.

Weed Management and Livestock

In mixed systems, where grass/clover leys are used for fertility building, livestock can make good use of the nutrients and they also produce manure, a resource which can be used around the farm to fertilise cash crops. Apart from leys, pastures will also need weed management and increasingly conservation of old or rough pastures requires specialist grazing. Animals can also be used to consume cut weeds or other plant material like chaff or screenings that are likely to contain some weed seeds.

Animals have different grazing habits and it is even recognised that different breeds or individuals are likely to have different tastes and habits. The species, breed, age and individuality of animals will all affect what they will eat and therefore what effect they will have on both weeds and pasture. Variability within the feeding site (e.g. vegetation, topography) can also be important as can other factors such as the weather. In general terms:

- Goats are browsers and have a reputation for enjoying tough and woody plants.
- Sheep are recognised as being useful for weed control as they graze close to the ground and will eat a wide range of plants. They can be used early and some breeds are hardy.
- Cattle can be used for early grazing but there are a large number of different breeds and types with different grazing requirements including beef, dairy and traditional breeds. Grazing strategies appear to be related to plant energy content and digestibility and this will affect how plants are eaten (leaves or stems or other parts of plants), which plants are eaten (species) and size plants eaten (young or more mature plants). Cattle tend to avoid longer coarser grass and hairy, spiny or poisonous plants. The selection of certain plant species and plant components as well as the location of these plants is based on the previous experience of the animals or learned from their mothers when they are calves.
- Pigs are good at rooting and have been recommended at various times for digging out perennial weeds like dock and couch when fenced within fields (and tightly stocked).
- Geese consume grassy weeds and have been used to weed in between rows of well established crops.
- Horses and ponies are grazed on ever increasing areas of land and can be used as part of a grazing rotation. They prefer frequent small amounts of fibrous grass or other high roughage material. They have been known to dramatically increase the number of docks in a rotation.

It is important to get the right grazing balance over the year to get the maximum benefit for the animals and also to prevent damage to the sward or soil. For example, stocking more lightly in the winter months and in wet periods prevents poaching. So think about the right season to graze, how long to graze, how many animals to graze and how long the grazing area will need to recover. Things to consider include:

- Timing grazing to benefit the pasture and promote competition with weeds.
- Timing grazing to damage the weeds, e.g. To remove flowers or seed heads before seed production.
- Allowing time for the pasture or forage to recover between grazings.
- Making sure that livestock that have been grazing on weedy land feed on weed seed free

forage for 4- 5 days before introducing them to weed free areas or pastures (some seeds will remain viable after passing through animals which may take a few days).

Suggestions for rotating livestock, depending on situation, include:

- Alternate the grazing of sheep and cattle from year to year or to use mixed grazing for better weed control. Mixed grazing in the same field may be detrimental to the cattle.
- Exploit animals.

Fallowing

It is usually not desirable to have to plan a fallow period into a rotation, but it may be necessary if weeds cannot be controlled during cropping or fertility building. It may not be necessary to stop cropping for a whole year, but instead to employ a bastard fallow i.e. no crop for part of the year. Tillage without a crop for a season is sometimes referred to as a black fallow. Fallowing the land for part of the growing season, as a bastard or summer fallow, can be as effective as a full fallow, is more suitable for lighter land and can be fitted into most rotations.

Fallowing is often best during the summer when cultivations can take place and the drier periods allow for root desiccation. This technique is more useful in plough-dominated systems rather than grassland management. One aim is to cultivate the soil progressively deeper over time, exposing underground plant parts to desiccation at the soil surface but in this case dry weather conditions are essential. Ploughing begins in June/July allowing time for an early crop to precede it. A bastard fallow is often used after a ley to reduce perennial weeds before sowing a winter cereal. There is also an opportunity for birds to feed on wireworms exposed during soil disturbance.

Fallowing has been shown to reduce perennial weeds within a rotation. The aim is to kill the vegetative organs of the weeds by mechanical damage and desiccation. For a full or bare fallow, heavy land is ploughed in April to give the weeds time to start into growth. It is cultivated or cross-ploughed 10-14 days later to produce a cloddy tilth. The soil is cultivated or ploughed at frequent intervals to move the clods around and dry them out. By August the clods should have broken down and the soil is left to allow the weed seeds to germinate. In September/October the weeds are ploughed in and the land prepared for autumn cropping. If a cereal is to follow the fallow, wheat bulb fly may be a problem because it lays eggs on bare ground in July. This can be overcome by sowing a green manure such as mustard to cover the land during this period.

Although there is the benefit of reduced weed control costs in subsequent crops after an effective fallow, the economics of taking land out of production for a full year together with undesirable effects on the soil and the environment, make the use of a bare fallow unlikely for weed control in the organic system. There is no financial return during the fallow period while labour costs accumulate during the fallowing operations. As an alternative to fallowing, cleaning crops such as potatoes and turnips allow repeated hoeing for weed control (but are not suited to heavy land).

A similar effect to that of fallowing can be achieved with rapidly developing crops like radish (Raphanus sativus) that are harvested before the onset of weed competition. The short interval between crop establishment and harvesting in this crop encourages weed seed germination but does not allow the weeds time to set seed or reproduce vegetatively.

Farm Hygiene

Weeds are, by their nature tenacious and almost impossible to eradicate once established. The best form of management is preventing their establishment in the first place. Weeds are easily spread between fields and between farms and it is worth taking some trouble to try and prevent this with some basic hygiene measures. Ask yourself the following questions:

Have you got a system for detecting weeds early?

- Managing a particular weed will be easier if it is detected early and prevented from spreading.
- Ensure that all people who work on the farm or visit it are alert to the possibility of spreading weeds and weed seed and ask them to tell you if they notice any particular areas of weeds.
- Keep records of problems weeds and their spread or otherwise. A digital camera can be a useful tool to record presence of weeds and monitor changes over time.

Is your farm machinery spreading weeds?

- Weed seeds are easily carried in soil, crop residues and on machinery so these should be regularly cleaned down.
- If there is a serious weed infestation in a particular field, or machinery is moving through fields where weeds are flowering, then washing down machinery should be a serious consideration.
- Hygiene is particularly important at harvest time. In crops like cereals weed seeds may be scattered in the field or caught on the machinery and dislodged later some distance from the original source. It may be necessary to add screens to combines to catch weed seeds at harvest. Older models may already have these features.

Tillage

Soil cultivation or tillage in its various forms has long been the mainstay of weed control and is the most effective way to reduce the weed seed bank. Seeds are encouraged to germinate and then the soil is cultivated mechanically to kill off the plants. The mouldboard plough is the traditional implement for burying weeds and crop residues as ground preparation for establishing a new crop. One piece of research showed that the annual loss of seeds from a natural soil weed seedbank (with no addition of fresh seed) was 22% with no cultivation. When the soil was cultivated twice a year the annual loss was 30%, and when cultivated four times it was 36%. However, it is not just the cultivations associated with the post-harvest incorporation of crop and weed residue that have weed control benefits. The method, depth, timing and frequency of cultivation may influence the composition, density and long-term persistence of the weed population.

Tillage is often divided into three types: primary, secondary and tertiary. However there are many operations that do not easily fall into these categories or span them all. We have provided an outline of the main tillage types below.

- Primary tillage: Is the principal cultivation operation before crop establishment. The main choice is between plough or non-plough systems (reduced or minimal cultivation, conservation tillage, no-till, direct-drilling etc. A range of machinery is available for primary tillage and some combinations can even work a stubble down to a seedbed in a single pass. Ploughing is seen as a method by which weed seeds can be buried below the depth from which they are capable of germinating, and it is sometimes said that ploughing can be used to bury a weed problem. This is particularly useful for small-seeded or annual grass weeds, which are often short-lived and may survive being buried. But this short-term solution to poor weed control in a previous crop often leads to long term problems due to the persistence of the buried weed seeds in the soil seedbank as viable seeds may be brought to the surface by ploughing in subsequent years and will germinate if conditions are suitable.
    - Reduced tillage: The concept of direct drilling crops without resorting to ploughing became popular after the development of the non-residual herbicides but recently, there has been renewed interest, primarily out of concern for soil conservation. Under reduced tillage there is better control of soil erosion, conservation of soil moisture and more efficient use of fossil fuel. However wind disseminated and perennial weed species can increase and volunteer weeds are also likely to be a problem. However, not all soils are suitable for reduced tillage and less nitrogen may be made available to crops where cultivation is reduced to a minimum.

Many organic farmers find it impossible to dispense with the plough routinely. Each method of cultivation has its advantages and disadvantages and the principle should be to use as many different types of cultivation as possible over the rotation which prevents any one type of weed getting the upper hand. Finer seedbeds produce more weed seedlings but a smooth surface makes direct weed control easier. Larger clods of soil produce fewer weed seedlings but the rough surface gives emerged weeds protection against direct weeding operations.

- Secondary tillage: Is used to prepare seedbeds and leave a level surface for drilling. Typically it involves disking or harrowing to a depth of 10 cm. Rotivators and power harrows are also used and are able to prepare seedbeds even when ploughing has not been carried out. Implements are available that can combine shallow seedbed preparations with some deeper cultivations in a single pass. Others can loosen the soil below the surface while leaving the preceding crop debris on the soil surface.
    - Stale or false seedbed: Is a seedbed prepared several days, weeks or even months before planting or transplanting a crop. The technique is recognised as a strategy suitable for organic farming and has been widely used for many years. The stale seedbed is based on the principle of flushing out weed seeds ready to germinate prior to the planting of the crop, depleting the seedbank in the surface layer of soil and reducing subsequent weed seedling emergence. The soil is cultivated about four weeks before drilling to stimulate germination and encourage the first, and usually biggest, flush of weeds. If irrigation is available it could be used to help.

- Tertiary tillage: Is the soil cultivation that is used directly as a means of physical weed control. It is dealt with in some detail under mechanical weed control.

Mechanical Weed Control

Mechanical weed control is mainly associated with cultivating tillage, often referred to as tertiary tillage, but also primary and secondary tillage as well as mowing and cutting have strong impacts on weeds. In organic farming, a 1–2-year period of grass-clover or alfalfa in the crop rotation is more or less required to suppress weeds including perennials such as Cirsium arvense. Perennial weed species such as C. arvense, Calystegia sepium and Agropyron repens are difficult to manage in crop rotations dominated by annual crops, and crop rotations dominated by cereals often result in a strong increase of root/rhizome biomass, especially if conservation tillage is practiced. A combination of repeated stubble tillage in the summer under dry soil conditions using a cultivator or rototiller combined with inversion tillage using the double-layer plough or a deep mouldboard plough may be successful in terms of weed control, but this control tactic can only be used occasionally in an organic cropping system, otherwise soil fertility will decline.

Cultivating tillage is carried out after crop sowing/planting to control weeds and consists of shallow tillage with a variety of equipments often categorized as hoes or harrows. It includes whole crop cultivation, inter-row cultivation and intra-row cultivation. The action mechanisms are through tearing weeds into pieces, uprooting and covering them with soil. The primary aim of cultivating tillage is to control seed propagated weed species in the earliest development stages. Perennial weeds with vegetative propagation are more or less unaffected by cultivating tillage.

Cultivating tillage is carried out both as pre-emergence and post-emergence cultivation. The primary control mechanism in pre-emergence cultivation is complex and little studied but in post-emergence weed harrowing, burying of weed seedlings in soil has been determined as the primary control mechanism. Plants should be covered totally to be killed; the required burial depth depends on plant size and growth habit.

Crop and weed populations are often not uniform in the field, which challenge the use and settings of cultivators. Weeds may occur in patches of varying size, densities and growth stages; some areas may have few or even no weeds within a weedy field. Also soil characteristics such as soil texture, soil moisture content and organic matter may vary significantly within a field. Therefore, there is need to vary the intensity (aggressiveness) of mechanical weed control according to the variations in the field. The objectives of site-specific weed management are to identify the variability, and to analyse and manage weeds according to their spatial and temporal variability. Lately, new technologies for sensing crops and weeds in real-time with image analysis, global positioning systems (GPS), mapping tools in a geographical information system (GIS) and robotics using autonomous vehicles allow a precise operation of mechanical weeding tools. This may increase the efficacy of weed control and reduce operation costs.

The objectives of this chapter are to give an overview of the various implements for mechanical weed control with their benefits and constraints. Secondly, this paper describes options to use existing mechanical weed control in a more precise way taking into account the heterogeneous weed distributions and soil conditions, i.e. under the precision weed management approach. In order to examine the potential of site-specific mechanical weed management, emphasis is given to cultivating tillage, which includes control of seed propagated weeds in early crop growth states.

Implements for Mechanical Weed Control

A large variety of implements are used for mechanical control of weeds, from basic hand tools to sophisticated tractor pulled or self-propelled implements. In general they are classified into two main groups: cultivating tools like hoes and harrows; and cutting tools like mowers and strimmers. For cultivating tillage only the first group is relevant.

Cultivating tillage may cover the crop row (intra-row weeding), strips between crop rows (inter-row weeding) or the full surface (whole crop weeding), and it is mainly carried out with harrows, hoes and brushes.

Inter-row and intra-row cultivation require precision in terms of steering. The highest precision is required for intra-row weeders but also inter-row weeders that operate close to the crop rows require precision. In practice, it is possible to leave about 10 cm wide uncultivated strips around the crop row if steering is highly accurate. Most inter-row cultivations are carried out in row crops with rows spaced 50–90 cm apart. Inter-row cultivation, however, is also possible in cereals and other crops normally established in narrow-row systems. Row spacing of about 20 cm is considered as a minimum to allow inter-row cultivation.

Harrows and Rotary Hoes

Weed harrowing and rotary hoeing imply whole crop cultivating and risks of crop damage. However, mechanical weeding may also favour crop growth due to soil loosening, reduction of evaporation, soil aeration and induction of mineralization. The challenge is to achieve a high degree of weed control without unacceptable crop damage. Spring tine harrows, also called flexible tine harrows, are the most utilized implements for whole crop cultivation in Europe, but other types of harrows like the chain harrow, are also used. In North America rotary hoes are commonly used pre-emergence or in very early growth stages.

(a) Whole crop cultivation in cereals with a 27 m wide spring-tine harrow, (b) Harrowing in faba beans.

Weed harrowing in cereals can be performed at pre-emergence, at early post-emergence, and during tillering until the crop becomes about 40–50 cm high; the latter is called selective harrowing. Early post-emergence harrowing is estimated to involve the highest risks for crop damage.

However, if broadleaved (annual) weeds are at cotyledon stage and the crop has reached at least three true leaves stage, weed harrowing can be highly selective due to higher crop resistance to cultivation. Under such conditions, 80–90% weed control may be achieved. In later growth stages, more aggressive cultivation is required which may involve several passes with the harrow, more aggressive setting of the tines or higher speed. Re-growth of weeds or late germinating weeds may require repeated cultivations, especially in row crops with open canopy structure and low competitive ability.

Harrowing is a common practice to control weeds in organic small-grain cereals and seed legumes. Early crop emergence, high competitive ability, and high tolerance to cultivation make cereals and seed legumes suitable for mechanical weed control. Weeds in maize (Zea mays L.), potato (Solanum tuberosum L.), and transplanted veg- etables may also be controlled by harrowing, and sensitive row crops like onions (Allium cepa L.), sugar beet (Beta vulgaris L.), and carrots (Daucus carota L.) can be harrowed after the 3–4-true-leaf stage of the crop.

Weather and soil conditions are important for successful weed harrowing. Relatively dry soil conditions are preferred, but there is little experimental evidence to quantify the importance of dry weather conditions. It is, however, important that mechanical weeding is not constrained by poor soil workability. Favourable soil conditions are crucial for the shallow tillage operations. Soil structure must be fine and soil texture light; stone free and plane topography are also desired. Harrowing on windy or hilly areas increases the risk of soil erosion. Additionally, harrowing in late autumn or early spring may provoke frost damage to the crop after treatment mainly due to root injury and reduced drought resistance.

The weed control mechanism of harrowing is mainly by soil burying (crop soil cover), but also uprooting plays a role when weeds are small. If cereals and seed legumes are damaged, it is mainly caused by crop soil cover. A high selectivity means that a high degree of weed control is associated with a small percentage of crop soil cover, which is the percentage of the crop that has been covered by soil. The success of harrowing depends on the balance between weed control and crop soil cover, which is reflected in the selectivity.

Originally, selectivity was defined as the ratio between percentage of weed control and the percentage of crop soil cover immediately after harrowing. This definition does not consider recovering or re-germination of weeds after treatment and until recently, crop soil cover was visually assessed due to lack of objective assessment methods. Recently, however, it has been possible to make reliable estimations of the crop soil cover based on digital image analysis, and a new protocol for the estimation of selectivity has been proposed.

It is important to distinguish between quantitative and qualitative aspects of selectivity, because the more aggressive the cultivation the lower selectivity. This applies for all implements and it is considered a quantitative aspect of selectivity. Therefore, comparisons of different treatments in terms of selectivity have to be done at the same level of weed control, in order to evaluate whether the treatments represent different qualities of selectivity. In general, the construction and use of cultivators are not considered to influence the qualitative aspects of selectivity whereas the difference between crop and weed plants (i.e. large crop plants and small weeds) is crucial.

Adjustment of the aggressiveness of harrows can be achieved by increasing the working depth (e.g. deeper penetration of the tines into the soil), the forward speed and the number of passes.

With the use of sensor technology and automatic plant species discrimination, crop and weed responses can be estimated and included in the objective determination of selectivity. The curve for leaf cover decline of the crop relative to increasing intensity with the cultivator allows the calculation of the percentage of crop soil cover, and the curve for weed density decline allows the calculation of the percentage of weed control. The selectivity is calculated as the relationship between crop soil cover and weed control.

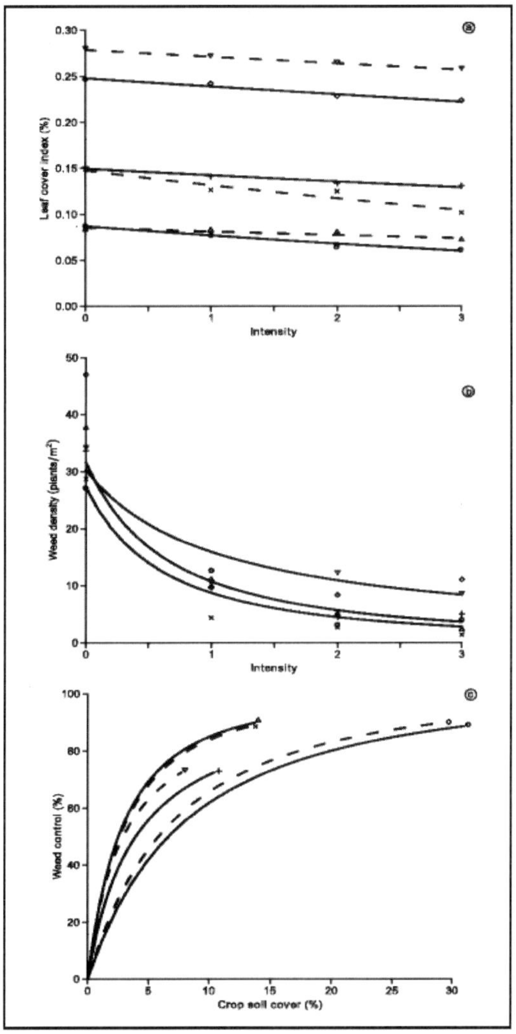

Leaf cover index (Symbols: ◦= BBCH 13, =BBCH 21, += BBCH 24) (a); weed density as function of increasing intensity (b); selectivity curves response of three timings and two driving directions of harrowing in spring barley (c). Solid lines indicate harrowing across and dashed lines along the crop rows. (Symbols b, c: = BBCH 13 along, ◦= BBCH 13 across, ×= BBCH 21 along, += BBCH 21 across, ∇= BBCH 24 along, ♦= BBCH 24 across).

After the data acquisition with the sensors, real-time adjustment of the aggres- siveness of cultivation (intensity) of the harrow may be possible, according to the site-specific demand for weed control. To make the adjustments, a predictive model is required to estimate the positive and negative crop yield contributions from harrowing relative to the intensity of cultivation. This model requires site-specific knowledge about (I) the competitive ability of the weeds, (II) selectivity and (III) recovery.

Prototype of an automatically controlled flexible tine harrow containing soil sensor (a), computing control unit (b), motor to change the tine angle (c), and RTK-DGPS for positioning (d).

The figure shows the prototype for automatic adjustment of the harrow. A digital sensor (a) measures the resistance force of the soil to the forward movement of the harrow; the data are transmitted to a control unit (b) that generates an algorithm for the motors (c) to change the tine angle and perform more aggressive treatments according to the soil characteristics. Positioning of the whole system is detected with a real-time kinematic differential global positioning system (RTK-DGPS) (d). The rotary hoe used for pre- and early post-emergence weed control without regard to crop rows, is a non-powered weeder with curved steel spokes radiating as a flat wheel from a hub. The spikes are rotated forward by ground contact.

Rotary hoes are mainly used in maize and soybean (Glycine max). They are less aggressive than most harrows, but have a high capacity due to high operation speed ranging from 8 to 24 km h^{-1}. The control mechanism is by pulling weed seedlings after the hoes penetrate and come up of the ground; also movement of the soil towards or away from the crop row generates a burial action. The selectivity of this tool is given when the crop seeds are placed deeper than its working depth – i.e. crop plants are deeper rooted than weed seedlings.

Inter-row Cultivation

Inter-row cultivators have been commonly used in row crops like sugar beets and vegetables for many decades. They are manufactured in many different designs, but they are all constructed to kill weeds on their path, which mean that they are operating in a non-selective way between rows. Most inter-row cultivators have sweeps, weed knives or shovels working in a depth of 2–4 cm. The ordinary hoe blades are mounted on rigid or vibrating shanks. Usually, three to five shanks constitute a gang that is munted on a toolbar, and each gang cultive an inter-row spacing. However, inter-row cultivation may also be carried out with rolling cultivators and PTO-driven cultivators.

As for harrowing, success of hoeing depends on dry weather conditions and a workable soil. Unlike harrows, hoes can be used rather in late growth stages and timing is not crucial. If the hoe is too deeply operated fibrous rooted weeds may grow again when enough moisture is available in the soil.

Instead of cutting blades, horizontal rotating brushes are used for special soil conditions. The weeds are brushed by rotation of hard polypropylene fibres and the control mechanisms are mainly by burial with soil and uprooting of weeds so they stay exposed to desiccation, stripping leafs and breaking stems. Manual guidance or autonomous guidance system of the brushes between the rows is indispensable. The first inter-row brush was developed in 1985 to be used in cereals with 17 cm row distance. For optimal weed control, the inter-row distance must be at least 17 cm. The main advantage of the brush weeders is that they can effectively be operated on higher soil moisture conditions than for harrows or hoes. The risk of using brushes is that soil structure is destroyed and the soil becomes very sensitive for compaction after rainfall.

Hoes are mainly used in row crops but may also be used in cereals as a supplement to whole crop cultivation with harrows. For instance in cereals, the effect of weed harrowing is often poor in heavy soils and a combination of inter-row hoeing and whole crop harrowing may improve weed management. Thus, an additional pass with the hoe might be more effective to control problematic weeds such as Galeopsis tetrahit, Galium aparine, Matricaria chamomilla and Vicia hirsuta. The crop row distance must not be <20 cm. Two passes with the hoe in maize and peas (Pisum sativum) may reduce inter-row weed density by 90% and intra-row density by 75%.

An accurate steering of the hoe is required, since its shares undercut everything when being pulled. Therefore precise seeding of the rows eases the guidance of the hoe between the rows. Manual steering has been the most common guidance method to direct the hoeing implements and reduce crop damage. Some electronic guided hoeing systems based on computer vision technology have been introduced. These systems aim to reduce the concentration needed by the tractor driver.

Intra-row Cultivation

There exist a number of implements for intra-row weeding. Most are low-tech, which means that they are simply pulled along the rows and the success of their performance is highly dependent on crop-weed selectivity factor. Among the most common low-tech implements are finger weeders and torsion weeders, which originate from North America but have been simplified by several companies.

The cycloid hoe is a high-tech device for intra-row cultivation. It was developed by the University of Osnabrück, Germany, together with Amazon Werke for weed control in maize. The machine had a multi-sensor system for plant recognition composed by three sensors: height-profile sensor, area-allocation sensor and soil-plant sensor. A cylindrical rotor works as actuator and contains eight tines placed around a vertical axis. The tines rotate in a circular motion, at a rotational diameter of 0.234 m. This translation movement of the rotor together with the forward straight-line movement of the implement generates a cycloid. The cyclic movement can be regulated by adjust- ing the translation and rotation speed. Every single tine can be in- and out-folded by an electromagnetic circuit to avoid crop plants, once the sensors have recognized them. The forward speed of the vehicle is 8.5 km h^{-1}.

The cycloid hoe has been further developed, tested and problems have been reported such as high crop damage and low control efficacy. Gobor designed a rotary intra-row hoe in combination with realt- time sensors for robotic weeding which is expected to be fast and effective in weed control.

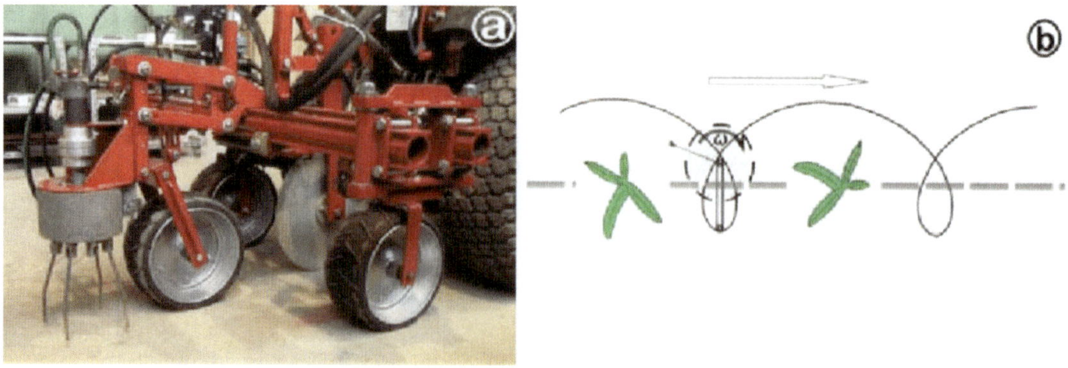

The cycloid hoe forward movement of the cycloid hoe.

Schweizer et al. carried out an investigation to perform more selective weed control through post-planting cultivating tillage with a so-called Bezzerides in-row cultivator, as an attempt to site-specifically manage weeds. The in-row cultivator has tools that move the soil away from the rows and later into rows, thus uprooting and burying in-row weeds. Rotary hoes at the first gang of the implement move soil away from the crop row in the first cultivation and into the row on the second cultivation, covering small weeds. The following gangs are composed by torsion weeders, spinners (rotary harrows), and spring hoe weeders. The torsion weeders and rotary harrows were used during the first pass; the torsion weeders and spring hoes (which replaced the spinners after the first cultivation) were used for the second and third passes.

Bezzerides-cultivator for in-row and intra-row weed control in maize.

Brush weeders also exist for intra-row weeding. However, they are designed with vertical brushes that are powered by hydraulic motors. The brushes can be assembled at any desired width and spacing for the crop; the working depth is about 20–30 mm.

Innovative Implements for In-row Crops

Innovative strategies for weed management in organic farming have been developed with several operating machines in Italy since the year 2000 in carrots, leaf-beet (Beta vulgaris var. cicla), fresh tomato (Solanum lycopersicum), and cabbage (Brassica oleracea). Weed control is carried out with a rolling harrow, a preci- sion hoe and flame weeders. All machines were built, tested and patented by the University of Pisa.

The rolling harrow was used for shallow tillage and showed efficient weed control. It is a modular machine, which means it can be built with different working depths to adapt it to the soil conditions.

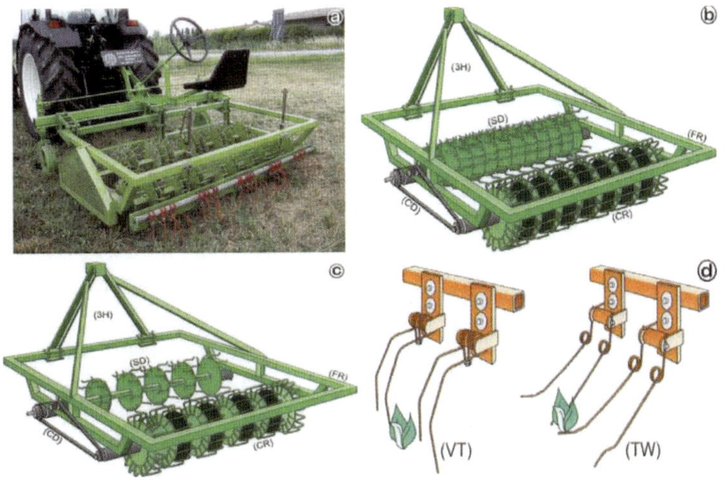

Rolling harrow (a) with a scheme of different working widths adaptation for whole crop cultivation (b), and inter-row cultivation (c). Parts: Three-point hitch (3H); frame (FR); front axle with spike discs (SD); rear axle with cage rolls (CR); chain drive (CD), (d) vibrating tines (VT) and torsion weeders (TW).

A square frame carries the working tools and has three points for linkage to the tractor. Spike discs of 30–35 cm diameter are placed in the front axle and cage rolls of 27–33 cm diameter placed in the rear axle; these axles are connected one another through a chain drive of an easily adjustable ratio.

The mode of action of the rolling harrow comprises the passage of the spike discs that till the soil at shallow layer of 4 cm; immediately, the gage rolls pass at a high peripheral speed as the rear axle is powered by the front axle, and generate an overdrive tilling and crumbling of the soil at a depth of 1–2 cm.

The discs and rolls may be arranged in a miscellaneous way and also changed with a simple blocking system. When the discs are narrowly distributed along the axle and the whole area is cultivated, a non-selective weed control for seed-bed preparation is obtained; whereas a wider arrangement of the discs allows an efficient selective inter-row weed control at post-emergence weeding. The minimum row distance to operate this machine is 15 cm.

Additionally, couples of elastic tines may be attached on the back of the frame (vibrating teeth and torsion weeders) in order to control weeds nearer or even in the crop row (inter-row control).

The precision hoe is a 2–3 m wide experimental inter-row weeder which shows promising results in organic horticulture in Italy.

This implement is structured on a square draw frame which bears up to 11 working tools each tool is placed on articulated parallelogram equipped with a small wheel that allows adjustment of the working width. The working tools consist of rigid elements which have attached a 9 cm wide triangular horizontal blade each, pairs of concave discs and two types of elastic tines: vibrating and torsion. The steerage is manually performed by a back-seated operator. With this machine it was possible to achieve a more selective weed control on the row in crops like cabbage, cauliflower and tomato.

Scheme of the precision hoe containing an operating seat (OS), steering handle (SH), articulated parallelogram (AP), directional wheel (DW), support wheel, lateral disc (LD), rigid elements with horizontal blade (HB) and elastic tines (ET) (a); precision hoeing in cabbage (b), in cauliflower (c), and in processing tomato (d).

Hand Weeding

The highest level of site-specific weed management is achieved through manual weeding with hand hoes, push hoes or simply through pulling weeds by hand. The major constraint is the low capacity. Nevertheless, hand weeding tools are still used in small-scale horticultural crops especially in organic farming. In crops like carrots and onions they often serve as the last resort control method.

Uprooting of perennial weeds in grassland is still carried out with hand tools such as the spudder which is very effective at removing deep rooted weeds.

Bond et al. describe some hand hoes used in Europe and still produced in a wide range of traditional and improved designs. Hoes with tough blades for cutting large weeds or with plough-shaped blades to move soil and cover small weeds are available. The draw hoe, swan-necked hoe, onion hoe, draw swan-necked, collinear hoe, Dutch hoe, Swiss oscillating, stirrup hoe are some examples. These tools cut and move the soil while the operator works standing or kneeling on vegetable crops. Their blades may be V-, L- or circle-shaped; they are mounted at right or slight angles to the short or large handles and are pushed, pulled or both to achieve the cutting action.

In developing countries, hand tools are of more readily acceptance and use due to low purchase costs. They are often used in combination with animal drawn hoeing or human pushing the wheeled hoe between the rows. In this way, manual weeding with hand tools is reduced only to intra-row level, reducing the labour per hour.

Cutting and Mowing

Cutting involves the mechanical removal of the above-ground portions of weed plants only, however plant damage by ordinary mechanical weeding on weed suitability for pest and diseases are also considered. This control method is better suitable for crops were the row distance is wide such as potato, maize, beets, and most of vegetables; cereals, rapeseed, peas and beans can be included when the row distance is > 16 cm.

Weed control by cutting or mowing may be complicated due to the adaptation mechanisms to continuous defoliation of some weeds, especially perennial ones in grasslands like Rumex obtusifolius, Cirsium arvense (L.) Scop and Pteridium aquilinum (L.) Kuhn. A high frequency of cutting of these weeds is required to achieve high degrees of weed control. Graglia et al. achieved about 75% weed biomass reduction of Cirsium arvense in spring barley, when 6 mowing per growing season were performed in preceding grass/clover leys.

In conclusion, mechanical weed control plays an important role in organic weed management; nevertheless conventional farming may also integrate mechanical weeding into its management plans. Efficacy and selectivity of mechanical weeders can be improved when variations in weed species distribution and soil characteristics are objectively assessed, and taken into account for the regulation of the instruments. Sensors to measure site-specific weed and crop responses to cultivation as well as soil resistance have been developed and combined with steerage and guidance systems. However, only few agronomic results are available showing the potentials of integrating advanced technology into mechanical weed control. So far, little is known about the interactions between soil characteristics in the upper soil layers, weed/crop densities, growth stages, weather conditions and mechanical weed control efficacy. Site-specific weed control may not only increase overall efficacy and selectivity but may also improve crop growth due to positive effects of soil tillage others than weed control. However, intensive studies need to be conducted to quantify positive and negative effects on crops from cultivating tillage to get decision support algorithms for precise mechanical weed control.

Physical Weed Control

Physical weeding methods have not yet been developed for highly selective control of weeds within crop rows. Thus, an area of 10 % to 20 % of e.g. a sugar beet (Beta vulgaris L.) field is either controlled by herbicide band spraying or by manual weeding. If this herbicide usage should be replaced or reduced, novel physical weeding methods have to be developed. Compared with traditional tractor pulled implements robotic systems show a different performance during the field operations. Light weight and low speed autonomous vehicles equipped with advanced sensor systems give new opportunities as they can work for 24 hours even under unfavorable field conditions.

Physical weeding methods suitable for robotic weeding in row crops are to be identified and selected. Crop:weed competition studies have shown that increasing distance between crop plant and weed seedlings increase yield significantly regardless of weed species. Thus, the physical weeding methods should be applicable in the within-row area including the area very close to the crop plants. The definition of the requirements and the performance properties is fundamental and a prerequisite for the decision about what weed control method should be used together with robotic weeding platforms.

To obtain efficient and accurate weed control systems they must be developed to target specific parts of the field. A field in which the crop is established in rows can be divided into three weeding target areas: (i) the area between the rows (inter-row area), (ii) the area between the crop seedlings within the rows (intra-row area), and (iii) the area close to and around the crop seedlings (close-to-crop area).

Inter-row hoeing, harrowing or brushing are matured methods within the last years and have reached a high level of automation even with automated guidance systems. These systems are economically viable for weeding in sugar beet.

For intra-row weeding new methods have to be developed for automated or autonomous vehicles. High accurate and active tools can be guided into the target areas within rows to allow tilling, cutting, pulling (pick) or just partly damaging the unwanted plants. Furthermore due to the ability of image processing to discriminate between species a high selectivity can be achieved. Selective weeding operations would only treat those weed plants that are regarded as harmful and to decrease crop yields. Others would then just be tolerated and could contribute to enhance the general biodiversity of the cropping system.

Objective

The investigation of physical weeding methods is a work package within an ongoing research project called 'Robotic Weeding' carried out in Denmark. The vision of the research project is to control weeds within rows and particularly close to the crop plants without damaging the crop plants. The specific objectives of this paper are:

- To define the requirements for physical weed control within crop rows.
- To review existing and possible future physical weeding methods and finally.
- To select methods suitable for further machine developments within an ongoing robotic weeding project.

Method

To decide about what weeding methods match best the particular requirements for weed control within rows the decision criteria have to be defined. The definitions of the criteria are based partly on a study of field conditions around individual sugar beet seedlings. Others are defined related to the demand of using small autonomous platforms to allow a high or even fully automated operation. The criteria to select physical weed control methods are listed below.

- Efficiency: Ability to control both annual and perennial weeds.
- Resolution: Spatial resolution or ability to target individual weed plants or even plant parts.
- Accessibility: To target weeds underneath crop leaves.
- Work rate: Treated weeds per time or area.
- Energy need: For the weed control operation.
- Applicability: Ability to be attached to an autonomous platform (size, weight, load etc.).
- Costs: Fixed and variable.

Requirements for Physical Weeding Close to Crop Area

The size of the close to crop area related to the intra-row area was found to be of importance. It was found that the size correlated with the developmental stage of crop roots and foliage. The radii of the foliage cover, the vertical and horizontal root growth distribution and the strength of crop seedling establishment define the border between intra-row and close to crop area. From the cotyledon to 8-leaf stage, an at least 6 cm radii around sugar beet plants allow no excessive soil disturbance, no leaf damage, require high resolution of height control above 2 cm. This is to avoid uprooting of crop roots. After the 8-leaf stage the sugar beet plants are more fixed in soil and therefore more tolerant to mechanical stressing.

Another finding from our study revealed information about where weed seedlings were located. Viewing sugar beet plants from above at the 6- to 8-leaf stage showed that the main shoot of weed seedlings to some extent and consistently were close to and hidden underneath beet leaves. Weeds close to the crop plants have a negative impact on crop yield. Therefore, an important requirement for a weeding method operating in the close to crop area is to be capable of targeting individual weeds underneath crop leaves.

To remove weed seedlings close to the stem of crop plants requires high control resolution. Furthermore, in some cases a protection of sugar beet leaf against damage has also to be obtained. The requirements for tillage depth in the within-row and close-to-crop area are different due to varying lateral and vertical root growth of sugar beet. There is a high risk of uprooting of whole sugar beet seedlings at very early growth stages within a radius of at least 2 cm from the plant centre. This is based on an estimate of the deviation between where a single sugar beet plant emerges related to its seed position influenced by soil type and seedbed conditions.

Intra-row Area

Considering the correlation of the size of the close-to-crop area and the growth stage the following requirements can be concluded for the intra-row area. Excessive soil disturbance is allowed and the requirement to the operational resolution is lower both for horizontal and vertical control. However, less target resolution implies higher operation distance to crop plants in order to minimise damage and uprooting.

As the intra-row area is the area outside the crop foliage cover makes the criteria for targeting weeds underneath beet leaves not relevant for the selection of methods for the intra-row area.

Selective Weeding

Highly selective control strategies shall successfully target individual weed seedlings during critical growth periods. Only controlling the key problem weed species minimizes the impact of weeds on crop yield and quality. Highly selective control strategies also minimise the overall treatment intensity and increase the work rate. Selective weeding requires high-resolution properties of the methods applied. The most high-resolution methods can treat individual weed seedlings. However, treating individual weed seedlings needs a reliable and applicable strategy in terms of defining species thresholds. The work rate can also be influenced by the selective weeding method because of its dependency on weed density.

The optimum timing for physical weed control is influenced by the first flushes of crop seedlings and weeds. The first and most numerous weed flush is emerging within 2-3 weeks after sowing, where the sugar beet seedlings have 2-4 leaves. The weed density also influences a second weed control, but in addition the competitive ability of the crop in the close-to-crop area and the growth stage of the weeds should be included in the weeding strategy.

Efficiency Requirements

Uprooting, cutting between stem and root or damage of main shoot can provide the physical control of most weed species. The most problematic weeds are the species that have their main shoot and leaves located close to ground level. These species can best be controlled by rupture of plant tissue at the main shoot or with a combination of shallow surface cutting and burial. Uprooting followed by burial provide the most efficient weed control of most annual and perennial weed species. Weed control efficiencies provided by principles different from uprooting followed by burial could provide good weed control by increasing the number of treatments and decreasing the time between each treatment.

Physical Weeding Methods

The process of selecting an appropriate physical weeding method for robotic weeding included a review of end effecters with properties that can provide the required weed control in the target areas. Some methods have already been applied to weed control and others are currently used in areas not related to agriculture. The physical weeding methods that in theory can be used for robotic weeding are listed in The methods listed are all direct weeding techniques for application at post-crop establishment. The weed control principles of each individual method can be categorized into four groups i) shoot:root separation, ii) uprooting, iii) burial and iv) plant tissue rupture.

Table: Physical weeding methods categorised by weed control principles.

Weed control principle	Physical method	End effecter
Shoot/Root separation	Cutting	Laser
		Water jet (+abrasive powder)
		Mower
		Vertical rotating disc (flail disc)
		Strimmers
		Narrow high pressure air jet stream with abrasive powder
	Plucking/Pulling	Grasp and twist weeder
Uprooting/Drying	Tillage	Goosefoot hoeing
		Horizontal rotating disc (concave disc)
		Brush weeder
		Torsion tines
		Vibrating steel rods
		Rotating tines (rotary or ‘cycloid hoe’)

		Soil engaging high pressure pneumatics
	Plucking/Pulling	Grasp and twist weeder below ground
Burial	Tillage	Spring tine harrowing
		Rotating tines (rotary or 'cycloid hoe')
		Rotor cultivator
		Brush weeder
	Cover (Mulch)	Mulching with various material
Plant tissue rupture	Electric	Electrical discharge
	Temperature (high)	Laser, micro-waves, UV-light, infra-red, hot water, steam
	Temperature (low) Pneumatic	Liquid N_2 High pressure air jet stream with abrasive powder

Selection of Weeding Methods

Results about weed control efficiency for the particular weeding methods identified have been difficult to find. Thus, the assessment and selection of the methods are based on the criteria mentioned, but in a more qualitative and depictive way.

Close-to-crop Area

The method of picking or pulling individual weed seedlings provides primarily a shoot:root separation, which does not provide the most efficient weed control of annual and perennial weed species. However, picking or pulling in combination with a device grabbing around and twist the weed seedlings in the upper 1 to 2 cm soil layer would provide the more efficient uprooting effect. The operation speed in terms of weed number removals per second is an important issue and has to be as high as possible. The method was chosen, because of the expected high weed control efficiency and ideal for selective weeding strategy. The applicability to an autonomous platform is good. However, the robotics controlling the end effecter needs to be very fast to enable this method to reach an acceptable work rate.

Lasers were found to be suitable for weed control due to a high-resolution performance and the ability of providing both a cutting and a thermal treatment. At the cotyledon to 2-leaf stage of weed seedlings the laser can be used as a thermal method. Thermal targeting of weed main shoots requires less energy than used for cutting the weeds. Probably at later growth stages, the laser can be used as a cutting tool, however with more energy requirements and with some difficulties in targeting weeds underneath leaves of sugar beet plants. All the methods belonging to the cutting category have one problem in common. They lack high control efficiency of weed species having their main shoot close to ground as explained previously. The energy requirements for lasers are high, especially when used for cutting. Nevertheless, the method was chosen because of its dual applicability, low weight and the possibility of high efficient thermal treatment of main shoot of weeds.

The last selected method provides both uprooting, cutting and burial at the same time. The method comprises a rotating horizontal disc (concave) of maximum 10 cm in diameter providing soil cultivation in the upper 1 to 2 cm soil layer. Therefore, the height control accuracy should be within sub-centimetre and the reaction time should be within milliseconds. The concave shape and rotation speed should be designed and defined to be able to operate in wet conditions and with stones.

The more shallow the operation, the less weed seeds germinate due to less soil disturbance. Even though the poor possibility of applying selective weeding strategies with this method, this method was selected because of the high efficiency of weed control and the simple end effecter. Depending on the robotics controlling the rotating disc, the control resolution of this method can be high.

It was found that not many of the selected methods were suitable to target weeds underneath sugar beet leaves. Thus, a guiding plate has to be structured to make the leaves of sugar beet slide to the side leaving the weeds applicable for any of the physical weeding methods. Another important finding was that for all methods height control with high resolution is required when operating in the close to crop area.

Intra-row Area

The horizontal rotating disc (concave) mentioned under methods selected for the close to crop area will also operate within the intra-row area because of its size.

The tilling rotary hoe ('cycloid hoe') was also chosen because of the uprooting effect and possibility of burial as well. These effects were found to be depending on the rotation and forward speed of the rotating tines. Replacing the straight with L-shaped tines enables this method to uproot weed seedlings underneath beet leaves without damaging beet leaves. Two effects are evident from a combination of one early physical weed control treatment in the close to crop area and a later treatment with the rotary hoe. Primarily, the overall work rate increases. Secondly, the weed control efficiency of combining these two methods can be high.

In conclusion, specific requirements for methods operating in the close to crop area are identified. This area requires targeting of weeds under beet leaves, no excessive soil disturbance and high control resolution both vertical and horizontal.

Many physical methods have properties useful for weed control in both the intra-row and close-to-crop area. The result of the selection of promising methods are:

- Grasp and twist weeder below ground (for close-to-crop and selective weeding).
- Laser for thermal treatment of main shoot of weeds (for close-to-crop and selective weeding).
- Horizontal rotating disc (concave) (for close-to-crop and intra-row weeding).
- Rotating tines/L-shaped tines (for intra-row weeding).

For the assessment and selection of the methods specific criteria were used based on field condition requirements and demands of using small autonomous platforms.

Chemical Weed Control

Herbicides are often the primary tools of choice for weed management across most acres of the Midwest. Many different herbicides and herbicide formulations are commercially available, including soil-applied and foliar-applied products, selective and nonselective products, products with long soil persistence, and products with no soil residual activity. The selection of which herbicide to use should be based on multiple factors, including soils, cropping rotations, tillage practices, and weed species. Sole dependence on herbicides may not necessarily provide the most eco- nomical or sustainable weed management. Integrating multiple practices reduces the likelihood of poor weed control due to unfavorable environmental conditions and reduces the intensity of selection for herbicide-resistant weeds.

Product Labels

Every herbicide product commercially available is required by law to have a label. The label provides a great deal of information about the product, including how it is to be applied, where, and in what quantity. The label is considered a legal document; using a herbicide in a manner inconsistent with its labeling is illegal. Herbicide labels change frequently, so be sure to consult the most current label when using a product. All pesticide products for sale in Illinois must be registered with the state government.

Application Rates

Herbicides applied at labeled rates should provide good weed control during the season of use while minimizing the potential for in-season crop injury and carryover into the following season. Herbicide application rates can vary according to many factors. Rates for soil-applied herbicides are greatly influenced by soil characteristics, such as organic matter content, texture, and pH. In general, heavy-textured soils high in organic matter often require a higher application rate than course-textured soils lower in organic matter. Application rates of postemergence herbicides are often determined by weed species and weed and crop size. For some postemergence products, higher application rates are suggested when certain weed species are present and when one or more weed species exceed a specified height or number of leaves.

Often several different commercially available formulations or premixes contain the same herbicide active ingredient. Much of the following text will demonstrate how to determine product equivalents and how to calculate amounts of active ingredient applied. Keep in mind that just because two or more products contain the same ingredient(s) does not necessarily mean they are applied at the same rates. Always consult the respective product label to determine the appropriate application rate.

Nomenclature

Across its lifetime a herbicide active ingredient may be sold by one or more companies and identified by one or more names. The three most common categories of names are trade, common, and chemical.

- Trade names: The trade name is the name under which a product is commercially sold; it is often the name most familiar to users. Examples of trade names include Valor, Raptor, Yukon, Basagran, and Cobra. These names are typically trademarked by the manufacturer so that no other company can use them. Trade names come and go, and sometimes they are recycled (for example, Option was once the trade name of a soybean herbicide but is now the trade name of a corn herbicide). You thus cannot always rely on the trade name to know what active ingredient(s) a product contains.

- Common names: Each common name is unique to a particular active ingredient. Common names are listed on the product label, usually in the active ingredient section. Flumioxazin, imazamox, halosulfuron plus dicamba, bentazon, and lactofen are the common names of the active ingredients contained in the commercial products Valor, Raptor, Yukon, Basagran, and Cobra, respectively. While more than one trade name may be used for a particular active ingredient, common names remain constant irrespective of trade names.
- Chemical names: Herbicide chemical names may not be as familiar as trade names or common names. Like common names, a chemical name is unique to a particular active ingredient, describing its chemical composition. For example, Salvo is the trade name of a herbicide with the active ingredient known by the common name 2,4-D, whose chemical name in turn is 2,4-dichlorophenoxyacetic acid.

Active Ingredients

The active ingredient of a pesticide formulation is the component responsible for its toxicity (phytotoxicity in the case of herbicides) or its ability to control the target pest. The active ingredient is always identified on the pesticide label, either by common name (for example, atrazine) or chemical name (for example, 2,4-dichlorphenoxyacetic acid). The active ingredient statement may also include information about how the product is formulated and the amount of active ingredient contained in a gallon or pound of formulated product. For example, the Basagran label indicates that the active ingredient (bentazon) is formulated as the sodium salt, and 1 gallon of Basagran contains 4 pounds active ingredient.

Usually when a herbicide trade name is followed by a number and letter designation (4L, 75DF, 7EC, etc.), the number indicates the pounds of active ingredient in a gallon (for liquid formulations) or a pound (for dry formulations) of the formulated product. So, for example, Basagran 4L contains 4 pounds of active ingredient (bentazon) per gallon of formulated product, AAtrex 90DF contains 0.90 pounds of active ingredient (atrazine) per pound of formulated product, and Prowl 3.3EC contains 3.3 pounds of active ingredient (pendimethalin) per gallon of formulated product.

Many herbicide labels restrict the maximum amount of product to be used per application and/or per year. These maximum rates are generally presented in terms of the total amount of active ingredient that can be applied per acre and/or per year. Several calculations can be used to determine the amount of active ingredient applied at a given product use rate. This is one of the easiest:

$$lb\ active\ ingredient\ applied\ per\ acre = \frac{gal\ or\ lb\ of\ product\ applied}{acre} \times \frac{lb\ active\ ingredient}{gal\ or\ lb\ of\ product}$$

So if we apply this equation to Basagran 4L, the amount of active ingredient (bentazon) applied at 2 pints (0.25 gallon) per acre of product is:

$$lb\ of\ bentazon\ (active\ ingredient)\ applied\ per\ acre = \frac{0.25\,gal\ of\ product\ applied}{acre} \times \frac{4\,lb\ active\ ingredient}{gal\ of\ product}$$
$$= 1\ lb\ active\ ingredient\ per\ acre$$

Types of Formulation

There are several ways to define formulation, but in essence it consists of the active ingredient and all associated components that make up the commercially available product. The active ingredient is responsible for controlling target weeds, but it rarely is the only component in a gallon or a pound of commercial herbicide. Other ingredients serve various functions, such as making the active ingredient safer and easier to handle, allowing the active ingredient to easily mix with water, and aiding herbicide uptake through plant leaves. These other components of a herbicide formulation are generally listed as inert ingredients on the product label, although they have important functions in making the active ingredient work as intended.

Several types of herbicide formulations are available, and a given herbicide active ingredient may be available in more than one formulation. Formulations are often designated on product labels as single or two-letter abbreviations. The more common herbicide formulations, along with their abbreviations, are presented in table.

Table: Common examples of herbicide formulations.

Type of formulation	Description of formulation
Flowable or aqueous suspension (F, L, or AS)	Liquid formulation containing finely ground solids suspended in a liquid.
Water-soluble concentrate	Liquid formulations that form a true solution when added to water.
Emulsifiable concentrate (EC or E)	Liquid formulation containing solvents and emulsifiers that disperse the active ingredient in water Dry formulation in which the active ingredient is sorbed onto aggregated granular particles.
Water-dispersible granule or dispersible granule (WDG or DG)	Dry formulation very similar to water-dispersible granules.
Dry flowable (DF)	A finely ground dry formulation (often mineral clays) onto which the active ingredient is sorbed.
Wettable powder (WP or W) Granule (G)	Dry formulation in which the active ingredient is coated onto an inert granule, ready to use without diluting in a liquid carrier.

- Acid equivalents: In some instances, the number preceding the formulation designation (L, EC, DF, etc.) indicates not pounds of active ingredient per gallon or pound, but rather acid equivalent per gallon or pound. Acid equivalent may be defined as that portion of a formulation (as in the case of 2,4-D ester, for example) that theoretically could be converted back to the corresponding or parent acid.

Another definition is the theoretical yield of parent acid from a pesticide active ingredient that has been formulated as a derivative (esters, salts, and amines are examples of derivatives). For example, the acid equivalents of the isooctyl and ethyl acetate ester formulations of 2,4-D are 66% and 88%, respectively. Why would a herbicide (one that has the acid as the parent molecule) be formulated as a derivative of the parent acid? An illustration using 2,4-D follows.

The herbicide active ingredient 2,4-D, originally discov- ered in the 1940s, continues to show utility across a diver- sity of landscapes. The herbicide is a popular tool among homeowners for selectively controlling certain broadleaf weed species in turf, and it is frequently a component of burndown herbicide applications in no-till agronomic cropping situations. Many commercially available 2,4-D formulations and trade names exist, but not all formula- tions and products are identical.

One characteristic of 2,4-D-containing products of particular importance is the type of formulation. Most often, 2,4-D products are available as one of three formulations: acid, amine, or ester. Each type has unique characteristics that can influence where and how a particular product is used.

The figure illustrates the chemical structure of 2,4-D. The molecule is considered a weak acid because the carboxyl hydrogen atom (the one to the far right) can dissociate, imparting a net negative charge to the molecule. In the dissociated (negatively charged) form, the acid molecule is very soluble in water but is not readily absorbed through a plant leaf. The waxy cuticle that covers the leaf surface is composed of many noncharged substances that reduce the ability of a charged molecule to penetrate and enter the plant. Somehow altering the parent acid form can influence how quickly and thoroughly it enters a plant through the leaf. These alterations produce derivatives that have physical and chemical properties different from the parent acid, such as increased ability to penetrate through a waxy leaf or increased water solubility for enhanced root uptake. The two most common derivatives of 2,4-D acid are amines and esters.

Esters are formed by reacting the parent acid with an alcohol, while amine salts are formed when the parent acid is reacted with an amine. The isooctyl ester is a very common ester formulation of 2,4-D, and the ammonium salt is perhaps the most common amine formulation. Other esters and amine salt formulations, however, are commercially available.

As previously mentioned, these different types of derivatives impart different characteristics to the formulation. For example, the isooctyl ester formulation is more soluble in hydrophobic ("water-avoiding") substances, like waxes, while amines are more soluble in hydrophilic ("waterloving") substances. In practical terms, esters are better able than amines to penetrate the waxy leaf surface of weeds, whereas amines are more easily moved into the soil by rainfall for root uptake (an important characteristic in certain brush-control applications).

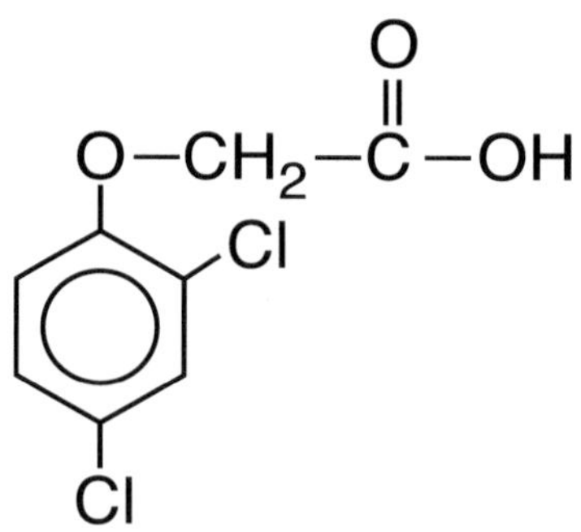

2,4 D parent acid.

Table provides some general comparisons between the amineand ester formulations of 2,4-D. These comparisons are somewhat relative since the specific type of amine salt or ester chain length can influence some characteristics. For example, ester formulations are considered more volatile (the change from a liquid state to a vapor state) than amine formulations, but the actual volatility potential of the ester formulation is influenced by the length of the ester chain (the number of carbon atoms). Also remember that different derivatives can impact the amount of active ingredient contained in a quantity of formulated product. To accurately compare among various products, calculations of "equivalency" should be based on the amount of acid equivalent contained in the formulation rather than the amount of active ingredient. An example follows of how to calculate acid equivalents, using ester formulations of 2,4-D as examples.

2,4-D can be formulated as various esters. The chain length of the ester can vary, but it is most commonly eight carbon atoms long (isooctyl ester). For this example, consider two ester formulations of 2,4-D: the first has only two carbon atoms forming the ester, and the second has eight carbons forming the ester. The parent acid is the same in these two formulations; the only difference is the length of the ester. These can be visualized in several diagrams.

The figure above illustrates the parent acid of 2,4-D; shows the parent acid formulated with a two-carbon side chain, and shows an eight-carbon side chain. While the carbon atoms of the side chain may modify some aspect of herbicide performance, it is the parent acid that acts at the target site within the plant. The additional carbon atoms of the ester side chain add weight to the formulation and may increase the amount of active ingredient of a formulation, but these atoms do not increase the amount of parent acid in the formulation. If these formulations were commercially available, and someone wanted to know how much of the parent acid each contained, the calculation would be based on the acid equivalents, not the active ingredients, of the formulations.

Table: Comparisons between amine and ester formulations of 2,4-D.

Amine salt	Ester
High water solubility.	Generally insoluble in water.
Low solubility in oils and waxes.	Higher solubility in oils and waxes.
Slow absorption into plant leaves.	Faster absorption into plant leaves.
No or very low volatility potential.	Low to high volatility potential.
Clear or slightly amber colored in water.	Milky when mixed in water.
Does not mix well with liquid fertilizers.	More compatible with liquid fertilizers.
Less preferred formulation for no-till burndown applications.	Preferred formulation for no-till burndown applications.
Reduced probability of crop injury following postemergence application.	Greater probability of crop injury following postemergence application.
Preferred formulation for in-crop (i.e., corn) applications when air temperatures exceed 85 °F.	Less preferred formulation for in-crop (i.e., corn) applications when air temperatures exceed 85 °F.

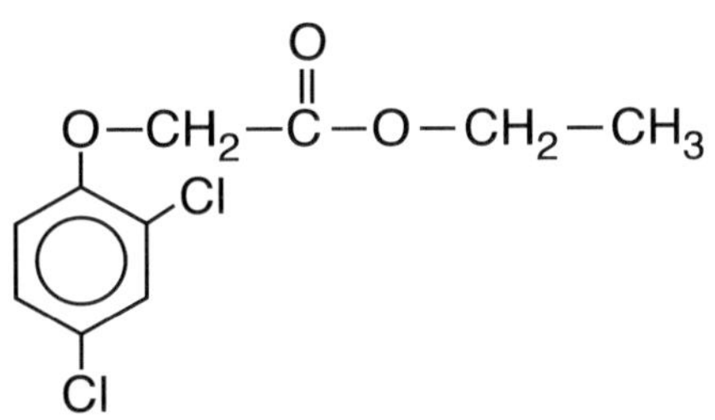

2,4-D ethyl acetate ester

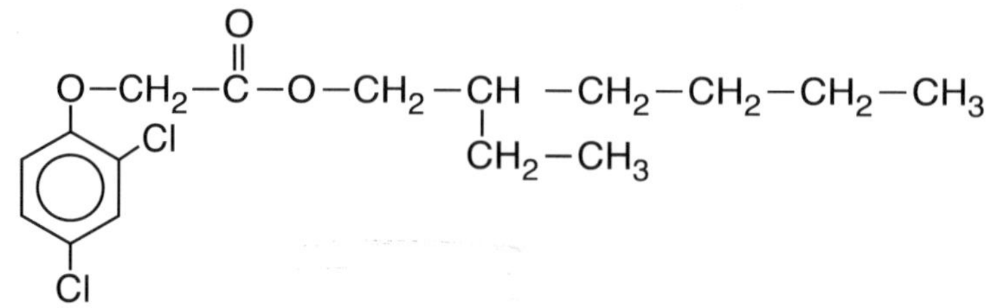

Isooctyl ester of 2,4 D.

Assume that both the two-carbon and eight-carbon ester formulations are commercially available and that each formulation contains 4 pounds of active ingredient per gallon. The application rate for both products is 1 pint per acre. Since the application rates and the pounds of active ingredient per gallon are identical for the two formulations, the amount of active ingredient applied would be the same for each. Verify this by performing the calculations previously illustrated for determining the amount of active ingredient applied. Even though the amounts of active ingredient applied are the same for the two formulations, the amounts of acid applied are not the same. Remember, it is the parent acid that binds to the target site to control the weed; the ester portion of the formulation is not involved in binding to the target site. What, then, is required to determine the amount of acid applied (i.e., the acid equivalent)?

The first step is to determine the amount of acid equivalent in a gallon of formulated product. Some labels indicate the amounts of both active ingredient and acid equivalent in a formulation, while others list only active ingredient. If the pounds acid equivalent is specified on the product label, all one need do to determine the pounds acid equivalent applied per acre is to substitute pounds acid equivalent for pounds active ingredient in the equation presented previously for calculating the pounds active ingredient applied. For this example, assume that neither 2,4-D label indicates the amount of acid equivalent.

The formula that can be used to calculate the amount of acid equivalent in a gallon of formulated product is:

$$\text{acid equivalent} = \frac{\text{molecular weight of the acid -1}}{\text{molecular weight of the salt or ester}} \times 100$$

Some molecular weights (i.e., how much the molecule weighs) are needed to complete these calculations. The molecular weight of the parent 2,4-D acid is 221.04. The molecular weight of the two-carbon ester formulation is 29.02 (weight of the two carbons and five hydrogens)+ 221.04 (weight of the parent acid) = 250.06. The molecular weight of the eight-carbon ester formulation is 333.25.

The acid equivalent of the two-carbon ester formulation is:

$$acidequivalent = \frac{221.04 - 1}{250.06} \times 100 = 88\%$$

tive ingredient per gallon, the amounts of active ingredient applied are equal. The amounts of acid (that part of the formulation that actually controls the weed) applied for each formulation are not equal.

The amount of acid applied per acre with the two-carbon ester formulation is:

$$lb\,of\ acid\ equivalent\ applied\ per\ acre = \frac{0.125 gal\ of\ product\ applied}{arce} \times \frac{3.52\,lb\,ae}{gel\ of\ product}$$
$$= .44\,lb\,ae\ per\ acre$$

The amount of acid applied per acre with the eight-carbon formulation is:

$$lb\,of\ acid\ equivalent\ applied\ per\ acre = \frac{0.125 gal\ of\ product\ applied}{arce} \times \frac{2.64\,lb\,ae}{gel\ of\ product}$$
$$= .33\,lb\,ae\ per\ acre$$

This example demonstrates that more 2,4-D acid is applied with the two-carbon ester formulation than with the eight-carbon formulation. In practical terms, more of the part of the formulation that actually controls the weeds was applied with the two-carbon ester formulation. To compare the herbicidally active portion of two ester, salt, or amine formulations, product equivalents should be calculated on the acid equivalent.

If only one formulation of a salt or ester product is commercially available, it wouldn't really matter if one calculated active ingredient or acid equivalent. For example, Pursuit is formulated as the ammonium salt of imazethapyr, but currently this is the only salt formulation commercially available for use in agronomic crops. There are, however, several commercial formulations of 2,4-D and glyphosate. Not all of these formulations contain the same amount of acid equivalent, so to determine equivalent rates among different formulations, calculations should be based on acid equivalent rather than active ingredient.

Since the commercialization of glyphosate-resistant soybean varieties in 1996, the number of glyphosate-containing products commercially available has increased dramatically. Currently, more than 50 such products are registered for use in Illinois agronomic crops, and that number is expected to continue increasing. Keeping track of product names and formulations can be daunting.

When selecting one of these products for weed control, keep several important considerations in mind: How much acid equivalent (ae) does the formulation contain? Should a spray additive (such as nonionic surfactant) be added to the tank, or does the formulation contain a "built-in" additive system? Are factors such as rain-free interval and toxicity category similar in the products you are consider- ing? Once these questions have been answered and you have narrowed down the list of products you're interested in purchasing, how can you compare costs? Should price comparisons be based simply on cost per gallon of formulated product? As in determining equivalent application rates, producers should compare prices on an acid equiva- lent basis.

To compare prices among glyphosate-containing products you need to do a few simple calculations. First, determine what rate to apply based on weed spectrum and size. For well-timed applications, a rate of 0.75 lb ae/acre can be very effective on many broadleaf and grass species. Once you have determined the application rate, calculate how many fluid ounces of each product are needed for this rate. Next, convert the price per gallon for each product to price per fluid ounce. Finally, multiply the number of fluid ounces needed to achieve the 0.75 lb ae/acre rate for each product by the cost per fluid ounce. An example to illustrate these calculations follows.

You decide to apply a glyphosate-containing product at 0.75 lb ae/acre when most broadleaf weeds are 4 to 6 inches tall. You are deciding between two glyphosate- containing products and want to know which offers the lowest cost per acre (for purposes of this example, assume additive requirements, if any are required by label, are identical for each product). "Glyfo A," a potassium salt, contains 4 lb ae per gallon and costs $23 a gallon. "Glyfo B," an isopropylamine salt, contains 3 lb ae per gallon and costs $21.75 a gallon.

Start by calculating how many fluid ounces are needed for an application rate of 0.75 lb ae/acre:

Glyfo A

$$\frac{0.75\, lb\, ae}{acre} \times \frac{1\, gal}{lb\, ae} \times \frac{128\, fl\, oz}{gal}$$

$$24\, fl\, oz$$

Glyfo B

$$\frac{0.75\, lb\, ae}{acre} \times \frac{1\, gal}{3\, lb\, ae} \times \frac{128\, fl\, oz}{gallon}$$

$$= 32\, fl\, oz$$

Next, divide the price per gallon by 128 to determine price per fluid ounce:

Glyfo A

$$\frac{\$23.00}{128\, fl\, oz} = \$0.1797/fl\, oz$$

Glyfo B

$$\frac{\$21.75}{128\, fl\, oz} = \$0.1699/fl\, oz$$

Finally, multiply cost per fluid ounce by the number of fluid ounces needed to achieve an application rate of 0.75 lb ae/acre:

Glyfo A

$$\frac{\$0.1797}{fl\, oz} \times \frac{24\, fl\, oz}{\text{acre}} = \$4.31/acre$$

Glyfo B

$$\frac{\$0.1699}{fl\, oz} \times \frac{32\, fl\, oz}{\text{acre}} = \$5.44/acre$$

So while a gallon of Glyfo A costs $1.25 more than a gallon of Glyfo B, calculating costs on an acid equivalent basis reveals that the per-acre cost is $1.13 less with Glyfo A than with Glyfo B.

Determining how many pounds of acid equivalent are contained in a given formulation may seem the most daunting part of this exercise, but several references are available that list the amount of acid equivalent in many commercially available glyphosate formulations. compares a number of glyphosate-containing products based on the amount of acid equivalent per gallon. The table also lists the amount of product (in fluid ounces) needed to apply a range of acid equivalents (0.375–1.5 lb per acre).

Table: Glyphosate-containing herbicides.

Active ingre- dent/gal	Acid equiva- lent/gal	Product rate equivalent to (lb acid equivalent/A)				
		0.375	0.56	0.75	1.13	1.5
		fl oz				
4	3	16	24	32	48	64
5	3.68	13	19.5	26	39	52
5.4	4	12	18	24	36	48
5.14	4.17	11.5	17	23	35	46
5.5	4.5	11	16	21	32	43
6.16	5	10	14	19	29	38

Herbicide isomers may not be very familiar to weed management practitioners, but they are becoming increasingly common in the marketplace. In essence, herbicide isomers are variations of a molecule, put together in slightly different arrangements. One isomer of a particular active ingredient is generally much more herbicidally active than the other isomer. A small amount of chemistry can help explain stereoisomers and how they are relevant in today's weed management arena.

A good starting point might be to define the term stereo- isomer. Stereoisomers are molecules that have the same atoms bonded to each other but differ in how the atoms are arranged in space. The figure illustrates a 5-carbon ring molecule with two chlo- rine atoms attached to it; one chlorine atom is positioned above the plane of the ring, while the other is positioned below shows the same 5-carbon ring with the same two chlorine atoms, but here both chlorine atoms are positioned above the plane of the ring. Each molecule contains the same number of atoms—5 carbon and 2 chlorine—but the spatial arrangement of the chlorine atoms differs, which is what differentiates this pair of stereoisomers. An analogy of stereoisomers is a person's two hands; each hand consists of the same components, but they are assembled differently. You cannot rotate your right hand to make it a left hand, and vice versa.

So how is a differential orientation of atoms or substituent groups (i.e., stereoisomers) relevant to weed management? Even though two molecules may have the same types and numbers of atoms and differ only in the orientation of one or more atoms or groups, differential orientations can greatly affect the biological activity of the molecules. If, for example, the molecules depicted in and were herbicides, the orientation of the chlorine atoms in might cause that isomer to bind much more effectively at the herbicide target site within the plant, whereas the orientation of the chlorine atoms in might not allow this isomer to bind the target site at all. One might reason that if the molecule depicted in is more herbicidally active than the molecule depicted in, it would be better to manufacture or use a product containing the molecule only. While this notion is valid, the process used to manufacture certain herbicides results in a combination of isomers (that is, a mixture of the two molecules) in the commercially available formulation. An example of stereoisomer chemistry in weed management is the active ingredient metolachlor.

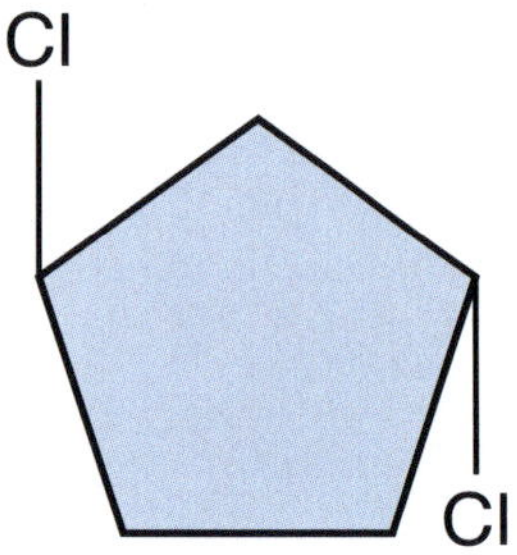

A 5-carbon ring with two chlorine atoms, one positioned above the plane of the ring and the other below.

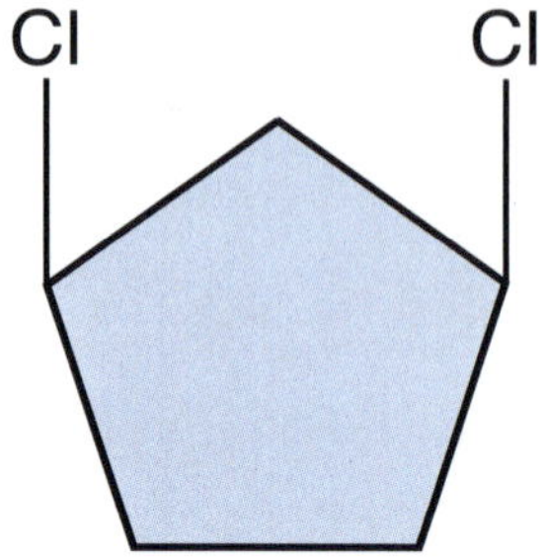

The same 5-carbon ring as shown in, but here both chlorine atoms are positioned above the plane of the ring.

Metolachlor first became commercially available during the 1970s and was sold under the trade name Dual. The process used to manufacture Dual resulted in two isomers of metolachlor present in the commercial formulation. One isomer, designated the S-isomer, is much more herbicidally active than the other, designated the R-isomer. Dual and the subsequent product Dual II each contained a 50:50 mixture of the active (S) and inactive (R) isomers of metolachlor. (Dual became Dual II when a safener was added to the original formulation to reduce the potential for adverse crop response.) Application rates for these "nonresolved" formulations were determined based on this 50:50 mixture of active and inactive isomers.

In the 1990s, improvements in technology allowed manufacturers to increase the amount of active (S) isomer in a formulation, and Dual II became Dual II Magnum. The "Magnum" formulations (Dual II Magnum, Bicep II Mag- num, Bicep Lite II Magnum) still contain the same active ingredient(s) as always, but they now contain a higher proportion of the active or resolved (S) isomer compared with the older formulations (Dual and Dual II, Bicep and Bicep II, Bicep Lite and Bicep Lite II). Specifically, the Magnum formulations contain an 88:12 mixture of the active (S):inactive (R) isomers compared with a 50:50 mixture of the active (S):inactive (R) isomers found in the Dual and Dual II formulations. So what is a practical implication of having a formulation containing more of the active isomer? Since a higher proportion of the active isomer is present in the Magnum formulations, application rates are reduced approximately 35% compared with the original formulation.

Perhaps another illustration will be of value. Say, hypothetically, you were to count out 100 molecules from a container of Dual II and 100 molecules from a container of Dual II Magnum. Assuming the rules of probability hold, the 100 molecules of Dual II would be 50 active molecules (the S or resolved isomer) and 50 inactive molecules (the R or unresolved isomer). The 100 molecules of Dual II Magnum would be 88 active and 12 inactive molecules.

Assuming the unresolved isomer doesn't contribute much to weed control, it takes less Dual II Magnum than either Dual or Dual II to obtain the critical number of S-metolachlor molecules needed for weed control. For example, if 50 molecules of S-metolachlor (the active isomer) are needed to achieve control of a particular weed species, how many total molecules of Dual/Dual II and Dual II Magnum would you need in order to apply at least 50 molecules of S-metolachlor? You would need 100 total molecules of Dual or Dual II (50:50 mixture) to get 50 molecules of S-metolachlor, whereas you would need only 57 total molecules of Dual II Magnum (88:12 mixture) to get 50 molecules of S-metolachlor. Stated another way, if you were to apply the same product rate of Dual and Dual II Magnum, you would apply less active isomer per acre from the Dual formulation.

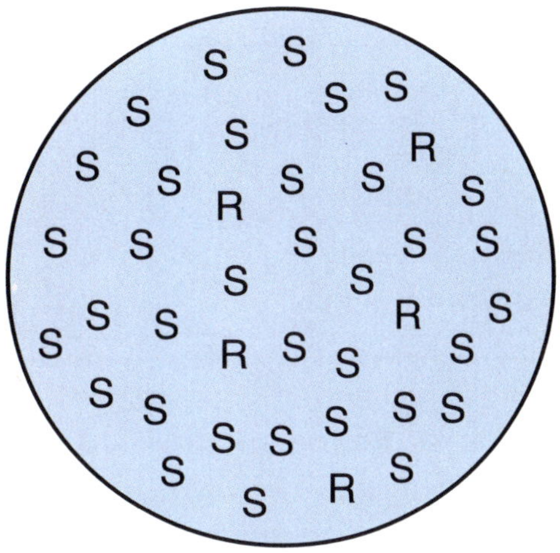

A droplet taken from a container of a non-resolved metolachlor-containing herbicide (50:50 mixture of S and R isomers). Note the equal numbers of S and R letters.

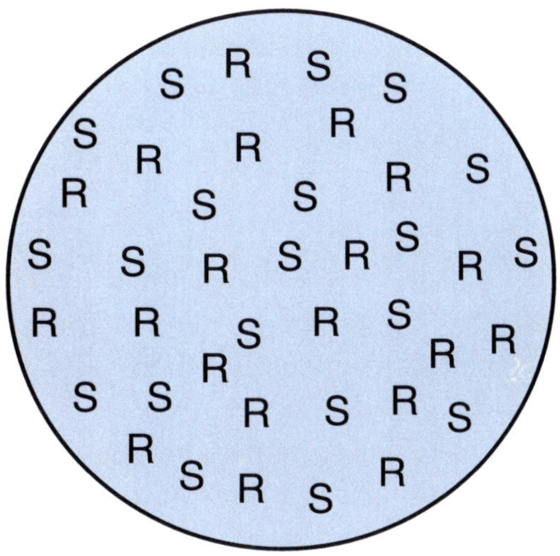

A droplet taken from a container of a re-solved metolachlor-containing herbicide (88:12 mixture of S and R isomers). Note the higher proportion of S letters relative to R letters.

The figures above illustrate this concept. The circles represent equal volumes of herbicide. Figure 12.13 was taken from a container of a nonresolved metolachlor-containing herbicide (50:50 mixture of S and R isomers) while was taken from a container of a resolved metolachlor formulation (88:12 mixture of S and R). Each circle contains the same number of total molecules (designated S and R), but a different proportion of S and R isomers.

This information should help those who purchase herbicides made up of stereoisomers better understand some of the differences among commercially available products. Currently there are many metolachlor and S-metolachlor products on the market, and there appears to be some confusion about product equivalents among these many formulations. For example, equivalent rates may be defined several ways, including equivalent amounts of active ingredients, equivalent amounts of active isomers, or simply the rates allowed by the respective product label. These are not always synonymous or interchangeable.

Table lists several examples of products containing metolachlor or S-metolachlor. One should not assume that applying the same rate of each product necessarily results in applying the same amount of active ingredient or active isomer. In particular, it should be noted that while applying the same product rates of an S-metolachlor-containing product and metolachlor-containing product can provide similar amounts of total active ingredient, the amounts of the active isomer applied can vary considerably.

Table: Metolachlor- and S-metolachlor-containing herbicides.

Product	Active ingredient	Active ingredient/ gal	R:S mixture (ratio)	If you apply (prouct/A):	You have applied	
					lb/ai	lb active isomer
Dual	metolachlor	8 lb	50:50	2.5 pt	2.5	1.25
Dual II	metolachlor	7.8 lb	50:50	2.5 pt	2.43	1.218
Dual Magnum	S-metolachlor	7.62 lb	88:12	1.67 pt	1.59	1.399
Dual II Magnum	S-metolachlor	7.64 lb	88:12	1.67 pt	1.59	1.403
“Generic I” brand	metolachlor	8 lb	50:50	1.67 pt	1.67	0.835
“Generic II” brand	metolachlor	7.8 lb	50:50	1.67 pt	1.62	0.814

Herbicide premixes are commercially formulated products containing more than one herbicide active ingredient. Combining two or more active ingredients in a formulated product can provide several advantages, including a broader weed control spectrum than any individual component has alone, reduced potential for physical or chemical incompatibility problems, and reduced cost compared with purchasing the components separately and mixing them.

Herbicide premixes can be confusing with respect to components, product equivalents, application rates, and other factors. Table below compares two commercially available corn herbicide premixes used in Illinois. The first column lists the trade name and formulation of the herbicide, and the second provides the common names for the components. For example, Harness Xtra (trade name) 5.6L (formulation) is composed of the active ingredients acetochlor (common name) and atrazine (common name). The second column also indicates the amount of active ingredient (or sometimes acid equivalent) of each component per gallon or pound of formulated product.

Table: Comparison of two herbicide premixes.

Herbicide	Components (ai/gal or lb)	If you apply/A:	You have applied (ai):	Product equivalents:
Bicep II Magnum 5.5L	S-metolachlor = 2.4 lb atrazine = 3.1 lb	2.1 qt	S-metolachlor = 1.26 lb atrazine = 1.63 lb	Dual II Magnum 7.64E = 1.32 pt AAtrex 4L = 3.26 pt
Harness Xtra 5.6L	acetochlor = 3.1 lb atrazine = 2.5 lb	2.5 qt	acetochlor = 1.94 lb atrazine = 1.56 lb	Harness 7E = 2.21 pt AAtrex 4L = 3.13 pt

The third column lists an application rate for each premix, and the fourth column indicates how much of each active ingredient is applied at that application rate. For example, 2.5 quarts of Harness Xtra 5.6L provides 1.94 lb acetochlor active ingredient and 1.56 lb atrazine active ingredient. Note here that while application rates of commercial products are usually expressed in ounces, pounds, pints, or quarts of product per acre, active ingredients are usually expressed in units of pounds active ingredient or acid equivalent per acre.

The last column lists product equivalents for each premix component when applied at the application rate listed in the third column. The 2.5-quart rate of Harness Xtra 5.6L provides the same amount of acetochlor and atrazine contained in 2.21 pints of Harness 7E and 3.13 pints of AAtrex 4L, respectively.

The application rate of Harness Xtra 5.6L listed in quarts per acre. Instead of 2.5 quarts, suppose someone would like to know how much acetochlor and atrazine are applied at a 2-quart rate of Harness Xtra 5.6L.

First, convert 2 quarts to gallons:

$$\frac{2qt}{acre} \times \frac{1\,gal}{4\,qt} = 0.5\,gal$$

Next, calculate how much acetochlor and atrazine active ingredient are contained in 0.5 gallon of Harness Xtra 5.6L.

$$\frac{0.5\,gal}{acre} \times \frac{3.1 lb\,ai\,acetochlor}{gal}$$
$$= 1.55\,lb\,ai\,acetochlor\ per\ acre$$

$$\frac{0.5\,gal}{acre} \times \frac{2.5\,lb\,ai\,acetochlor}{gal} = 1.25\,lb\,ai\,acetochlor\ per\ acre$$

Finally, determine product equivalents based on these active ingredient amounts:

$$\frac{1.55\,lb\,ai\,acetochlor}{acre} \times \frac{1\,gal\,Harness}{7\,lb\,ai} \times \frac{8\,pt}{gal} = 1.77\,pt\,Harness\,7E$$

$$\frac{1.25\,lb\,ai\,acetochlor}{acre} \times \frac{1\,gal\,Harness}{4\,lb\,ai} \times \frac{8\,pt}{gal} = 2.5\,pt\,AAtrex\,4L$$

Principles of Soil-applied Herbicides

Soil-applied herbicides remain an important part of weed management programs in corn and soybean production systems. Early preplant (EPP), preplant incorporated (PPI), and preemergence (PRE) surface are the most common types of herbicide applications to soil. EPP applications are typically made several weeks prior to planting and are more common in corn fields than soybean fields. PPI applications were once very common, but they have declined in recent years with the growing adoption of conservation tillage. PRE applications are generally made within one week of crop planting. Regardless of when or how a herbicide is applied to the soil, the effectiveness of soil-applied herbicides is influenced by several factors.

For a soil-applied herbicide to be effective, it needs to be available for uptake by the weed seedling (usually before the seedling emerges, but some soil-applied herbicides can control small emerged weeds under certain conditions). Processes such as herbicide adsorption to soil colloids or organic matter can reduce the amount of herbicide available for weed absorption. Soil-applied herbicides do not prevent weed seed germination; rather, they are first absorbed by the root or shoot of the seedling and then exert their phytotoxic effect. Generally, this happens before the seedling emerges from the soil. For a herbicide to be absorbed by weed seedlings, the herbicide must be in the soil solution or vapor phase (i.e., an available form). How is this achieved? The most common methods for herbicides to become dissolved into the soil solution are by mechanical incorporation or precipitation. EPP applications in no-till systems attempt to increase the likelihood that sufficient precipitation will be received before planting to incorporate the herbicide. If, however, no precipitation is received between application and planting, mechanical incorporation (where feasible) will, in most instances, adequately move the herbicide into the soil solution. Herbicide that remains on a dry soil surface after application may not provide much effective weed control and is subject to various dissipation processes, some of them described in subsequent paragraphs.

Many weed species, in particular small-seeded ones, germinate from fairly shallow depths in the soil. The top 1 to 2 inches of soil is the primary zone of weed seed germination and should thus be the target area for herbicide placement. Shallow incorporation can be achieved by mechanical methods or precipitation. Which method is more consistent? Precipitation provides for fairly uniform incorporation, but mechanical incorporation reduces the absolute dependence on receiving timely precipitation.

How much precipitation is needed and how soon after application it should be received for optimal herbicide performance depends on many factors, but generally 1/2 to 1 inch of rain within 7 to 10 days is sufficient.

Herbicides remaining on the soil surface, or those placed too deeply in the soil, may not be intercepted by the emerging weed seedlings. Herbicides on the soil surface are subjected to several processes that reduce their availability. Volatility (the change from a liquid to gaseous state) and photolysis (degradation due to absorption of sunlight) are two common processes that can reduce the availability of herbicides remaining on the soil surface. Volatility potential is determined by several properties of the soil and the herbicide formulation, while photolysis is dependent primarily on herbicide properties.

Dry soil conditions are conducive for planting, but they may also reduce the effectiveness of soil-applied herbicides. If herbicide applications are made prior to planting and no precipitation is received between application and planting, a shallow mechanical incorporation prior to planting may help preserve much of the herbicide's effectiveness.

Principles of Postemergence Herbicides

Postemergence herbicides are a key part of an integrated weed management program. Applications made after crops and weeds have emerged allow for identifying the weed species present and assessing the severity of infestation so that herbicide selection can be tailored to the particular field. Postemergence herbicide applications minimize the interactions of the herbicide with factors associated with soil (such as texture and organic matter content), but they often magnify interactions between the herbicide and prevailing environmental conditions.

To achieve weed control with postemergence herbicides, the herbicide must come in contact with the target, be retained on the leaf surface prior to absorption into the plant, be able to reach the site of action within the plant, and finally induce some phytotoxic response. If for any reason one or more of these steps is restricted or limited, the level of weed control can be expected to decline.

The plant cuticle serves as an outer protective layer, or "barrier," that restricts the amount of water lost by the plant through transpiration. It also serves a variety of other functions, and the cuticle is often considered the primary barrier that limits herbicide absorption. The cuticle is composed primarily of waxes and cutin, substances that effectively limit water movement out of the leaf (transpiration) or into it (absorption). The type and amount of wax that comprises the cuticle influences the degree of wetting that can be achieved, and this composition can change with plant age and in response to changes in the environment. Older plants and plants under environmental stress generally have more wax or a different structure of the wax comprising their cuticles and are thus more difficult to wet. One of the main functions of certain spray additives is to enhance herbicide penetration through the cuticle.

Plant age and size, relative humidity, soil moisture, and temperature are other factors that influence absorption of postemergence herbicides. Younger, smaller plants usually absorb herbicide more rapidly than older, more mature plants. Many postemergence herbicide labels recommend applications be made when target weeds are small and caution about reduced effectiveness if applications are made to larger plants. Labels of postemergence herbicides may also suggest that users delay applications if weeds are under "adverse environmental conditions." Examples of such adverse environmental conditions may include prolonged periods without significant precipitation (resulting in dry soil) or low air temperatures. On the other hand, high relative humidity, adequate soil moisture, and moderate to warm air temperatures all favor enhanced herbicide absorption. Remember that if conditions occur that enhance herbicide absorption into weeds, conditions also are favorable for enhanced absorption into the crop, which may result in crop injury.

Post emergence herbicides vary in their mobility within the plant. Some demonstrate very limited movement following absorption and are commonly referred to as "contact" herbicides. Others can move extensively within the vascular elements of the plant and are referred to as "translocated" herbicides. Contact herbicides do show some movement following absorption, but they do not move nearly as exten-sively as translocated herbicides. Thorough spray coverageof the plant foliage is very important with contact herbicides but somewhat less important with translocated herbicides.

Almost every post emergence herbicide has a pre harvest interval specified on the label or a crop developmental stage beyond which applications should not be made. Labels of some products indicate both a developmental stage and a pre harvest interval. A pre harvest interval indicates the amount of time that must elapse between herbicide application and crop harvest. Such intervals are established to allow sufficient time for the herbicide to be broken down or metabolized in the plant. Additionally, the preharvest interval reduces the likelihood of herbicide residue remaining on the harvested portion of the crop. Failure to observe the pre harvest interval may result in herbicide residue in the crop in excess of established limits. In addition to pre harvest intervals, there are restrictions on many post emergence herbicides labels about whether the treated crop may be used for livestock feed or whether treated fields may be grazed as forage.

Another interval that is important to observe is the rotational crop interval. Nearly all herbicide labels, both soil-applied and post emergence, list rotational crop intervals that specify the time that must elapse between herbicide application and planting a rotational crop. This becomes particularly important with late-season herbicide applications. Such intervals are established to reduce the likelihood that sufficient herbicide residues will persist in the soil that could adversely affect the rotational crop. Some herbicide rotational restrictions are based solely on time, while others are influenced by different factors, such as soil pH and the amount of precipitation received after herbicide application.

Additives for post emergence herbicides. Additives are compounds added to a herbicide formulation or spray mixture that in some way modify the characteristics of the spray solution. Additives either are included in the commercial herbicide formulation or are added to the spray mixture prior to application. Different types of spray additives perform different functions, such as improving herbicide uptake into the target vegetation, reducing the number of very small droplets so as to reduce physical drift, and enhancing herbicide performance on certain weed species. Some of the most common additives for postemergence herbicides are nonionic surfactants (NIS), crop oil concentrates (COC), and ammonium fertilizer salts. These are used to increase the effect of the herbicide on the target plants:

- Nonionic surfactants lower the surface tension of spray droplets, thus increasing spray coverage, so they are frequently referred to as spreaders or wetting agents.
- Herbicide labels often specify that the NIS should contain a minimum of 75% to 80% active ingredient or otherwise use a higher rate of NIS. NIS is usually applied at 0.5 to 1 pint per acre, or 0.125% to 0.5% on a volume basis.
- Ammonium fertilizer adjuvants are added to increase herbicide activity on certain weed species, including velvetleaf. The two most common ammonium fertilizers used are ammonium sulfate (AMS) and urea ammonium nitrate (UAN) solution (28-0-0). AMS is used at 8.5 to 17 pounds per 100 gallons of spray solution. UAN is used at 2 to 4 quarts per acre, or 2% to 4% by volume. Contact herbicide labels may specify that fertilizer adjuvants replace NIS or COC, while translocated herbicides often specify the addi- tion of UAN or AMS to NIS or COC.
- Crop oil concentrates are phytobland oils with emulsifiers added to allow mixing with water. The oil may be of petroleum (POC) or vegetable (VOC) origin. Oils increase spray penetration through the leaf cuticle. Most herbicide labels allow POC or VOC, but some may specify one or the other only. COCs are used at 1 to 2 pints per acre, or 0.5% to 1% by volume.

- Compatibility agents are spray additives that improve mixing, especially for soil-residual herbicides that are applied with a liquid fertilizer spray carrier. Herbicide labels often specify a "jar test" to determine the need for a compatibility agent when mixing herbicides with liquid fertilizer. The rate is usually 1 to 4 pints per 100 gallons of spray mix.
- Drift reduction agents are added to the spray tank to reduce small droplet formation and thus minimize drift potential. The use rate per 100 gallons of spray is generally 2 to 10 fluid ounces of concentrated forms and 2 to 4 quarts of dilute forms (1% to 2% active ingredient).
- Buffer-surfactants or buffer-compatibility agents contain organic phosphatic acids that provide an acidifying effect on spray mixes where a pesticide is affected by alkaline water. Most herbicides do not need a buffering agent, and some sulfonylureas should not be acidified because herbicide degradation is accelerated.

How Herbicides Work

Herbicides are frequently categorized into families according to various similarities. Examples of classification categories include mode of action, application timing, and chemical structure. Herbicide mode of action describes the metabolic or physiological plant process impaired or inhibited by the herbicide. Essentially, mode of action refers to how the herbicide acts to inhibit plant growth.

Herbicide site of action describes the specific location(s) within the plant where the herbicide binds. Site of action thus identifies the herbicide target site within the plant. Though the most common herbicide classification schemes utilize mode of action, much ambiguity exists with respect to that herbicide classification.

While understanding herbicide mode of action is beneficial, classifying herbicides by site of action may be more useful from the standpoint of resistance management.

Herbicide resistance in plants is often due to an alteration of the binding site in the target plant. Rotating herbicides based on these different binding sites or sites of action may provide for more reliable classification, in contrast with the ambiguity of classification based on herbicide mode of action, whose systems include anywhere from seven to 13 different categories. Some of these systems describe mode of action categories as "cell membrane disruptors," "seedling growth inhibitors," and "amino acid synthesis inhibitors." Rotating herbicides based on these categories could cause confusion among growers. For example, the mode-of-action category "amino acid synthesis inhibitors" would place the herbicides Pursuit (imazethapyr) and Roundup (glyphosate) in the same family, whereas classification by site of action would place these two herbicides into two distinctly different families, allowing growers to more accurately rotate herbicides for resistance management.

Biological Weed Control

Biological control agents usually target their specific natural enemy weeds. Recently due to certain favorable environmental, health, economic and sustainability reasons; foreign and native organisms that attack weeds are being evaluated for use as biological control agents that may be used to complement conventional methods especially where some weeds have developed resistance to chemical control. Wheeler et al. reported that their international team discovered and tested numerous new species of potential biological control agents that could attack different plant tissues such as defoliators, sap-suckers, stem borers, and leaf- and stem-gall formers. Many successful biological weed control programs in many parts of the world have demonstrated the potency of this approach and support the concept that natural enemies can contribute to the reduction of plant growth and reproduction. Wapshere et al. classified biological approach to weed control as follows: the classical or inoculative method which is based on the introduction of host-specific exotic natural enemies adapted to exotic weeds; the inundative or augmentative method which is based on the mass production and release of native natural enemies usually against native weeds; the conservative method which is based on reducing numbers of native parasites, predators and diseases of native phytophages that feed on native plants; and the broad-spectrum method which is based on the artificial manipulation of the natural enemy population so that the level of attack on the weed is restricted to achieve the desired level of control. According to McFadyen classical method is the predominant method in weed biocontrol. He further explained that classical method involves the introduction and release of agents in form of exotic insects, mites or pathogens to give permanent control, while inundative involves the releases of predators, use of bioherbicides and other integrated pest management which usually are not as widely used as the classical method. Also there are three different techniques for applied biocontrol: (i) conservation—protection or maintenance of existing populations of biocontrol agents; (ii) augmentation—regular action to increase populations of biocontrol agents, either by periodic releases or by environmental manipulation; and (iii) classical biocontrol—the importation and release of exotic biocontrol agents, with the expectation that the agents will become established and further releases will not be necessary.

Louda and Masters stated that despite the positive impact of chemical herbicides in agricultural productivity, complete reliance on chemical control has caused severe problems such as high cost per unit area, decreasing effectiveness, negative impact on plant diversity and increased environmental contamination. He therefore pointed out that the use of biological factors that naturally limit weed populations is one promising alternative. Menaria discussed bioherbicides as an eco-friendly approach to weed management. He explained that the use of chemical herbicides leaves some chemical residues in food commodities which directly or indirectly affect human health. According to him this situation led to the search for alternative methods that are environmentally friendly, and biocontrol has been found a suitable alternative. Green reviewed the potential for control using bioherbicides of four important forest weed species in the UK; including bracken, bramble, Japanese knotweed and rhododendron. They concluded that rhododendron is a suitable target weed for control using wood-rotting fungus as a bioherbicide stump treatment; and this is an approach already developed for weedy hardwood species in South Africa, Canada and Netherlands. Clewley et al. analyzed factors associated with control programs (invasive region, native region, plant growth form, target longevity, control agent guild, taxonomy and study duration) in order to identify patterns of control success. They found out that biological control agents significantly reduced plant size $(28\pm4\%)$, plant mass $(37\pm4\%)$,flower and seed production $(35\pm13$ and $42\pm9\%$, respectively$)$ and target plant density $(56\pm7\%)$.

Underlying Principles and Procedures for Biological Weed Control

Underlying Principles

The underlying principle behind biological approach to weed control is based on some research works that reported that exotic plants become invasive because they have escaped from the insect herbivores and other natural enemies that limit their multiplication and distribution in their native regions; however some other factors may contribute to the tendency for particular plant species to become invasive. Therefore biological control involves using specific natural enemies that can diminish the development and reproduction of their prey organism and put some limitations to them. McFadyen stated that the predominant approach to classical biological weed control involves the importation, colonization, and establishment of exotic natural enemies (predators, parasites, and pathogens) to diminish and maintain exotic pest populations to densities that are economically insignificant.

General Procedures

Some authors have outlined general procedures to be followed when embarking on classical biological weed control programs as follows: (i) evaluate the ecology, economic impact of the weed and potential conflicts of interest; (ii) survey the organisms that are already attacking the weed in the new habitat in order to distinguish accidentally introduced agents and so eliminate such from future evaluation; (iii) carry out literature search and other forms of survey to identify natural enemies attacking the weed in its native region; (iv) screen the possible biological control agents in the foreign country to determine host range and specificity, and to remove nonspecific agents from further consideration; (v) carry out further tests of promising candidates in quarantine after introduction to ensure host specificity and eliminate predators, parasites, and pathogens that may have been introduced with them; (vi) embark on mass rearing of host-specific agents; (vii) release the host-specific agents; (viii) carry out post-release evaluation to determine establishment and effectiveness of agents; and (ix) redistribute agents to other areas where control is required. Wapshere et al. presented a summary of steps normally followed when introducing a biological control agent in a classical biological control weed program as in table below.

Steps	Details
1. Initiation	Data on taxonomy, biology, ecology, economics, native and introduced distributions, known natural enemies, etc., are compiled by initiating scientist or group. An extensive literature review is conducted on the proposed target weed and its relatives, plus known natural enemies. Conflicts of interest identified and resolved if possible.
2. Target weed approval	Data in (step 1) submitted to appropriate State and Federal groups for comment; additional data may be required.
3. Foreign exploration and domestic surveys	If project approved in (step 2), the center of evolution of the genus of the target weed (if known) and other suitable areas, are searched for natural enemies, particularly where these are eco-climatically similar to the area of introduction. At the same time, the weed should be investigated in the country of introduction for attacking enemies, related plants, etc.
4. Weed ecology and agent host specificity	Ecology of the target weed, its close relatives and its natural enemies is studied in the native area, and the most damaging and apparently selective agents are subjected to several years of host-specificity testing.
5. Agent approval	A report on each agent is submitted to appropriate State and Federal bodies to obtain importation and release permits.

6. Importation and quarantine clearance	Each agent is imported to the country of introduction where it is reared through at least one generation in quarantine to rid it of its parasites and diseases.
7. Rearing and release	After a pure culture of the agent is obtained in (step 6), it is normally mass-reared and released in the field in cages or free at field sites.
8. Evaluation and monitoring	Agent is monitored at field sites to determine establishment and degree of stress on target weed, or to determine reasons why the agent did not become established or efficacious.
9. Redistribution	To aid spontaneous self-dissemination, agent is distributed to other areas in the target weed's distribution, if needed.

Reasons for Relatively Slow Popularity and Adoption of Biological Weed Control

Recent research activities and weed control practices around the world have shown that the old idea derived from untested opinions; that biological approach to weed control is usually very slow, unpredictable, expensive and mostly unsuccessful is totally not true. Apart from the high initial costs, biological approach to weed control has been known to be relatively cheaper when compared to other methods; however certain factors have slowed down the rate of adoption. These factors include: long time of establishment-usually 20 years or more to ensure success, inadequate or no records of the extent of pre-biological control weed infestations that should serve as a guide for a new biocontrol program, discouraging story of poorly implemented weed bio-control programs. A lot of success stories however have been documented. Lack of information about previous successfully implemented biological control of weeds often lead to untested theories becoming established dogma and this negatively influence the decisions to or not apply it. For instance Mcfadyen stated that it was believed that biological control of trees is difficult, but many examples of trees controlled by insects have been reported. Also classical biological control has been viewed as unsuitable for weeds of annual crops or other frequently disturbed environments, however there are many examples of successful control of crop weeds.

Some researchers have reported that there are evidences showing that some agents introduced for exotic weed control have attacked non target, native plants; and this situation has raised concerns among biological control workers and weed scientists as well as the governments. Opposition to biological approach to control of weeds has also contributed to slowing down the rate of adoption and practice; this is because some researchers and weed control scientists believe that it is difficult to estimate the cost or the feasibility of biocontrol. Based on a study carried out in South Africa, it was reported that some of the weed biocontrol projects have provided practical solutions to problems e.g. the development of Stumpout for the treatment of wattle stumps and the use of C. gloeosporioides for the control of H. sericea. However other projects have been less successful and have resulted in the rejection of potential agents for various reasons and these include C. albofundus on A. mearnsii, X. campestris on M. aquaticum and G. nitens on R. cuneifolius. Vurro and Evans identified legislative hurdles, technological and commercial constraints as limitations to the adoption of biological weed control in Europe. Olckers stated that limited budgets in many countries have also helped to slow than the rate of adoption and practice of biological approach to weed control.

Examples of Successful Biological Control of Weeds

One thousand one hundred and forty-four individuals (mostly entomologists and plant pathologists) have ever attended the International Symposia on Biological Control of Weeds (ISBCWs), and out of these, 450–550 weed biological control experts have been actively involved in research and development efforts over the last 50 years mainly from USA, Canada, Australia, South Africa and New Zealand. McFadyen reported that biological approach to weed control has a long history and a good success rate of 94. A comprehensive list of agents and their target weeds have been documented by Winston et al. Culliney presented potential benefits estimated for some proposed or initiated biological control programs targeting invasive weeds. Frequently cited examples of successful approach to biological weed control are the prickly pear cacti (Opuntia; spp.) in Australia, eradicated by an imported moth (Cactoblastis cactorum) and rangeland in California, Oregon, Washington, and British Columbia controlled by St. John's wort Hypericum perforatum (millepertuis perforé). Mcfadyen presented a list of 41 weds which have successfully been controlled using introduced insects and pathogens and another three weeds also controlled by introduced fungi applied as mycoherbicides. He further stated that many of these successes have been repeated in other countries and continents. Julien presented a list of both successful and failed cases of biological weed control; this included the introduction of 225 organisms against 111 weed species, and 178 insects and 6 mites. Palmer et al. reported that 43 new arthropod or pathogen agents were released in 19 projects; and that effective biological control was achieved in several projects with the outstanding successes being the control of rubber vine, Cryptostegia grandiflora, and bridal creeper, Asparagus asparagoides.

Determination of the Success of Weed Biological Control

Information collated on weed impacts before the initiation of a biological control program is necessary to provide baseline data and devise performance criteria with which the program can subsequently be evaluated. For avoidance of confusion on when a biological control could be viewed as successful or not, Hoffmann stated that an implementation of a particular biological control will be termed successful when: complete-when no other control method is required or used, at least in areas where the agent(s) is established; substantial-where other methods are needed but the effort required is reduced (e.g. less herbicide or less frequent application); and negligible-where despite damage inflicted by agents, control of the weed is still dependent on other control measures. Complete control does not imply total eradication of the weed; rather it means that control measures are not required anymore specifically against the target weed, and that crop or pasture yield losses will not be attributed mainly to this weed. Substantial control involves situations where control may be complete in some seasons and/or over part of the weed's range, as well as cases where the control achieved is widespread and economically significant but the weed is still a major problem. It is therefore concluded that successful implementation of biological approach to weed control is the successful control of the weed, and not necessarily the successful establishment of individual agents released against the weed. Successful biological control depends on three factors: the extent to which each individual agent can limit the targeted plant; the ecology of the agent as it affects its ability to populate and spread easily in the new environment; and the ecology of the weed, which determines if the total damage that can be caused by the agent can significantly reduce its population. Because agents always need some surviving predator plants to complete their life cycle, biological control will not usually totally eradicate their target weeds. In essence a successful biological control program reduces the potency and population of the target weed and usually in conjunction with other control methods as part of an overall integrated weed management scheme which is recommended.

Making the Choice of Agents to be Introduced to Control Weeds

Gassmann reported that selection of potential agents in the last decades has been mainly based on the population biology of the weed, impact studies of agents on the plant and the combined effect of herbivory and plant competition. Palmer et al. stated that agent selection is highly dependent on the type of weed, its reproductive system, on the ecological, abiotic and management context in which that weed occurs, and on the acceptable goals and impact thresholds required of a biological control program. Generally, factors to be considered in selecting agents include the following: the agent must target a particular plant species, must have high level of predation and parasitism on the host plant and its entire population, must be prolific, must be able to thrive in all habitats and climates where the weed exists and should be able to spread easily and widely, must be a strong colonizer, the overall cost of introducing the agent must be cheaper compared to other control methods, the technology that will be involved in introducing and managing the agent must be as simple as possible, must as much as possible maintain natural biodiversity, sufficient number of individuals must be released, plant phenology (effect of periodic plant life cycle events) must be favorable. To be considered a good candidate for biological control, a weed should be non-native, present in numbers and densities greater than in its native range and numerous enough to cause environmental or economic damage, the weed should also be present over a broad geographic range, have few or no redeeming or beneficial qualities, have taxonomic characteristics sufficiently distinct from those of economically important and native plant. Furthermore, the weed should occur in relatively undisturbed areas to allow for the establishment of biological control agents, cultivation, mowing and other disturbances can have a destructive effect on many arthropod bio-control agents. Inundative biocontrol agents such as bacteria and fungi are less sensitive to these types of disturbances so may be used in cropland.

Steps to Identifying and Introducing Biological Control Agents

The study of insect attributes and fitness traits, the influence of plant resources on insect performance, and the construction of comparative life-tables, are the first steps towards an improvement of the success rate of biological weed control. Generally, steps to identifying and introducing biological control agents include: (i) identify target weeds; (ii) identify control agents and determine the level of specialization; (iii) apply controlled release of the agents; (iv) apply full release and determine optimal release sites; (v) for the case of classical methods, monitor release sites; (vi) apply redistribution for the case of classical methods (vii) and maintain control agent populations.

References

- Factors-in-cropping, agricultural-technology: britannica.com, Retrieved 14 January, 2019
- Weedman, crops: omafra.gov.on.ca, Retrieved 7 March, 2019
- Cultural-weed-controls: gardenorganic.org.uk, Retrieved 19 May, 2019
- Mechanical-Weed-Control: researchgate.net, Retrieved 2 February, 2019
- Biological-methods-of-weed-control, biological-approaches-for-controlling-weeds: intechopen.com, Retrieved 29 June, 2019

Chapter 3

Herbicides: Mechanisms of Action

The chemical substances which are used to control unwanted plant or weeds are known as herbicides. The classification of herbicides is done on the basis of mechanism of action or site of action. Mechanism of action refers to the first protein, enzyme or biochemical step which is affected in a plant when the herbicide is applied. All the diverse aspects of herbicides and their mechanism of action have been carefully analyzed in this chapter.

Herbicides

Since the introduction of agriculture, humans have battled weeds in an effort to reduce crop and livestock losses. Physical control of weeds such as hand weeding and primitive forms of tillage dominated early weed control. The Romans first introduced chemical weed control (approx. 300 BC) when they applied salt and olive oil to control weeds in crops and along roads after they noted that the ground became barren beneath their olive oil presses. In the 19th and 20th Centuries inorganic compounds such as sulphuric acid, copper and iron sulphate, lead arsenate, copper nitrate and sodium arsenate were used to control broadleaf weeds in cereal crops. In 1880 sprayers were developed to deliver a mist of herbicides.It was not until the 20th Century that the production of effective and selective herbicides was recorded. Synthetic herbicides (e.g., 2,4-D, MCPA) were developed during World War II and eventually first marketed for weed control in 1944. These synthetic auxins revolutionised weed control and gave companies the impetus to research and develop the large array of herbicides on the market today.

Triazine herbicides ushered in a new era of pre- emergent weed control in maize and other row crops, as well as vineyards and orchards. Simazine was the first triazine to be used commercially in 1956. It was developed by J. R. Geigy in Switzerland and approved for use on maize, asparagus, grape rootstocks and rights-of-way. In 1958, the herbicide atrazine, also developed by Geigy, was first registered for weed control in maize in the USA. Atrazine was, and still is, an extremely successful herbicide because it is broad spectrum, low cost, and allows flexible timing of applications. Even today, with more than 60 other herbicides registered for maize in the USA, more than two thirds of the maize crop is treated with atrazine. Other herbicide modes of action followed and today there are over 300 herbicide active ingredients across 27 modes of action. However, of these 300 active ingredients 79% fall into only eight herbicide modes of action.

Herbicide Development

Agricultural companies now spend U.S.$ 50 to 200 million to develop and register each new herbicide. Herbicides are heavily regulated in most countries and frequently need to obtain approvals from many different regulatory agencies. The primary government agencies responsible for herbicide regulation in most countries are environmental protection agencies, departments of agriculture and the food and drug agencies. New herbicides must be registered with the appropriate regulatory authority in each country. This requires a wide range of tests and a thorough safety and efficacy review prior to obtaining registration of a new product. Herbicides must be registered for use on different crops, with each registration requiring a range of safety information. The costs for herbicide registration are extensive and increase with each crop for which the herbicide will be registered.

Discovery

Novel compounds must be synthesised, put through initial screening in the laboratory or greenhouse,patented, and analysed. A great deal of cost is expended to determine the physio-chemical properties of the active ingredient molecules (environmental fate, weed/crop selectivity, metabolism studies, etc.).

Formulation

The product must be formulated so that it will be stable under a wide range of conditions, can be readily taken up by plants, and does not present significant exposure risks to end-users and the environment. The product also must be able to be manufactured cost effectively.

Toxicology

Herbicide registration requires a battery of toxicological tests that are analysed and submitted to regulatory agencies for review. These agencies require a wide range of base studies that include acute and long-term effects, chemical properties, effects on endangered species, fate in the environment, persistence, etc.

Marketing

Once the product is close to registration companies must begin planning a product launch campaign. This includes acquiring and training personnel for sales and marketing, and developing an advertising campaign for the new product.

Major Crops Targeted

For economic reasons, companies pursue herbicides that have utility in high volume crops, such as maize, wheat, soybean, and rice in order to help recoup the high expense of herbicide development. If the herbicide has been successfully launched in one of these high volume crops then eventually companies may pursue registration of their new product in minor crops or for non-crop situations.

Herbicide Application

Herbicides can be applied at a number of stages through the cropping cycle. The most common application times are given in the adjacent text box. Optimal application times are determined by the type of crop and the type of herbicide.

Herbicide Application Times

Pre-plant

The herbicide is applied onto the bare soil surface and/or onto emerged weeds prior to planting. Pre-plant herbicides are often broad-spectrum herbicides, able to control a wide range of plants.

Pre-plant Incorporated

The herbicide is applied prior to planting and is incorporated into the soil to reduce herbicide loss by volatilisation or photo-degradation.

Pre-emergence

Pre-emergence herbicides are applied to the soil after planting but prior to the emergence of the crop. These herbicides generally require rainfall or irrigation to move the herbicide into the soil for maximum activity.

At Cracking

These herbicides are applied as the crop emerges or "cracks" the soil surface. Selectivity is gained by the fact that there is limited uptake of the herbicide in the terminal bud of the crop.

Post-emergence

Post-emergence herbicides are applied over the top of the crop and the weeds. The crop must have sufficient tolerance to the herbicide for this practice to work. Post-emergence herbicides are generally applied soon after weed emergence, as larger weeds are more difficult to control. Most herbicide tolerant crops are sprayed post-emergence.

Directed Post-emergence

Some crops, such as cotton, have woody stems that limit uptake of herbicides. This allows farmers to direct herbicides at the weeds at the base of the crop, avoiding the higher foliage and growing tips of the crop.

Shielded (Hooded) Post-emergence

Hooded sprayers may be used to hold herbicide sprays within the hooded area and so shield crops from non-selective herbicides during spraying.

Mode and Mechanism of Action of Herbicides

Herbicides kill plants by disrupting an essential physiological process. This is accomplished by the herbicide specifically binding to a single protein for many herbicides. The target protein is referred to as the herbicide "site of action." Herbicides in the same family generally have the same site of action. The symptoms observed on weeds sprayed with herbicides express the mode of action (MOA). In other words, the mechanism by which a herbicide kills a plant is known as its "mode of action (MOA)". For example, triazine herbicides interfere with photosynthesis by binding to the D_1 protein involved in photosynthetic electron transfer. Thus, the site of action for triazines is the D_1 protein, whereas the mode of action is the disruption of photosynthesis. An understanding of herbicide mode of action is essential for diagnosing crop injury or off-target injury problems and for designing weed management programs with a low risk of selecting for herbicide-resistant weed populations. Knowing a herbicide MOA is important to understand how to use a herbicide most effectively. In herbicide R&D, when new chemicals are screened experts carefully observe the detail and timing of the appearance of symptoms to gain clues as to the MOA. Fully understanding a MOA may take years of research by plant physiologists, biochemists, molecular biologists and many other scientific disciplines. The more that is known about MOA, the more safely and effectively herbicides can be used.

Mode of Action

Most often, herbicides have MOAs which mean that a target enzyme no longer functions properly, or at all. This is usually because the herbicide molecule has distorted the enzyme molecule in some way. Shape is crucial to enzyme function. Enzymes are 'catalysts'. They provide a platform for a specific chemical reaction which they are not directly part of and which would not happen effectively without them. 'Knocking-out' an enzyme with a herbicide has two main effects with various consequences:

- Component chemicals for the reaction accumulate and may be directly or indirectly damaging.
- Absence of the reaction's product will restrict growth, either through starvation of key building blocks or because the reaction makes chemicals which normally protect the plant.

As plants have evolved, slight differences in their biochemical systems for carrying-out physiological processes have arisen. Sometimes a gene with DNA coding for a particular enzyme or other herbicide target will have mutated in some species so that variations in that enzyme exist which may differ by only one or two amino acids out of thousands. This small change will still allow plants to grow successfully, but it may mean that herbicides with a MOA involving this enzyme will only be effective on one type and, therefore, on a particular group of species.

Gene mutations are responsible for one sort of herbicide selectivity based on MOA. Herbicides such as fluazifop-p-butyl inhibit the enzyme Acetyl Coenzyme A Carboxylase (ACC-ase) and selectively control grass weeds in broad-leaved crops like soybeans or oilseed rape (canola). The enzyme carrying out the same function in broadleaved weeds and crops is slightly different and not affected by the herbicide. On the other hand, glyphosate is a non-selective herbicide with a very wide weed control spectrum because it inhibits the enzyme 5-enolpyruvylshikimate-3-phosphate (EPSP) synthase, a fundamentally important enzyme very similar in all plants.

Mutations are also one way in which weed resistance to herbicides can occur. Rare individual plants which have a mutated enzyme may not be susceptible to a herbicide which is generally effective on that species. An outbreak of weed resistance occurs when such an individual is allowed to reproduce and comes to dominate the weed population in a field. Ensuring that herbicides with different MOAs are used helps to avoid weeds becoming resistant.

The other main type of selectivity and weed resistance occurs when one group of plants can metabolize a herbicide rapidly enough to avoid any damage.

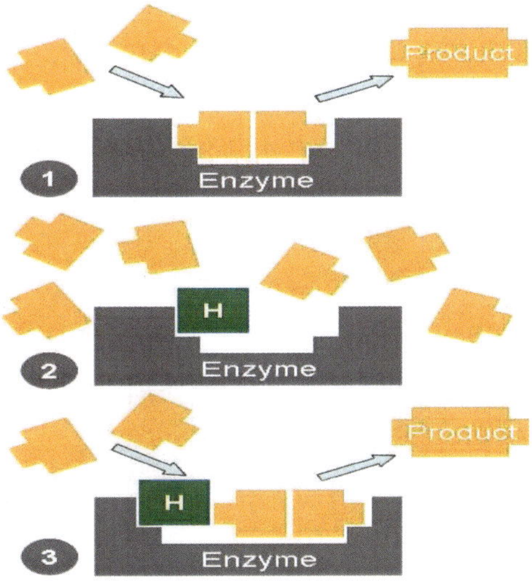

The figure above shows an enzyme functioning normally. In 2) a herbicide is stopping biochemicals coming together to form the usual product. As a result the plant is starved of the product and potentially damaging building blocks accumulate. 3) shows an enzyme which has mutated so that even when the herbicide is present the reaction can continue. Either the herbicide is selective to species with this type of enzyme, or an individual plant with this mutation could reproduce to establish a population of resistant weeds.

Several other types of herbicide MOA exist. Some affect photosynthesis in various ways to ultimately divert the flow of energy from sunlight. Instead of being incorporated into chemical energy stores it ends up in very destructive 'free radicals' based on oxygen. These destroy membranes and desiccate leaves. Another large group, the first selective herbicides to be discovered, mimic plant hormones – the chemical messengers that control growth and development. Another group affect cell division by, for instance, disrupting the mechanism by which chromosomes are arranged in a cell as it prepares to divide in two. Others herbicides have more general chemical effects which are hard describe as specific MOAs, or where very little has been discovered about the MOA. These are generally older products which have been superseded.

Classifying Modes of Action

Herbicide MOAs can be classified by the plant process affected, eg photosynthesis, cell division or specific enzyme targets. This often reflects the practical uses of herbicide products. For instance, those affecting photosynthesis are applied post-emergence and are often fast acting; those affecting cell division are often pre- emergence herbicides affecting germinating seeds; and some enzyme targets will determine patterns of weed control spectrum and selectivity depending on whether the enzyme is fundamental to the growth of all plants or whether it is confined to, say, grasses or broadleaved weeds.

However, there are always significant exceptions to these general rules and their particular chemistry may mean that they move in plants and soil differently. These factors determine selectivity, whether they are contact or systemic herbicides, and whether they can be used pre-emergence, post-emergence, or both. A more detailed classification is required for managing weed resistance.

MOAs Involving Photosynthesis

Four types of herbicides primarily affect photosynthesis. Paraquat and its sister herbicide diquat are the only members of the first type. When the energy in sunlight is captured by chlorophyll it is transferred in a flow of electrons through 'Photosystem I'. Paraquat diverts this flow resulting in the production of highly reactive free radicals which very quickly destroy cell membranes, spilling the contents, and appearing as yellowing and desiccation. This happens within hours in bright sunlight because of the high levels of energy running rampant. Almost all green plants are affected by paraquat making it a broad-spectrum, non-selective herbicide.

The second type blocks the transfer of energy through 'Photosystem II'. Herbicides of this type ('PS II inhibitors') bind to a protein involved in the transfer chain, reducing its effectiveness. This causes some diversion of electrons with similar, but slower, results to paraquat and slows growth because of less energy for photosynthesis. Triazines like atrazine and ureas like chlortoluron are both PSII inhibitors, but have slightly different MOAs because they bind to the protein concerned in different places.

A third type, the protoporphorinogen oxidase (PPO) inhibitors, interfere with an enzyme involved in producing chlorophyll and other large molecules important in photosynthesis. Without new chlorophyll leaves yellow and photosynthesis slows. However, unused building blocks of chlorophyll accumulate and react with oxygen to form reactive free radicals which are especially destructive to broadleaved species.

The fourth type of photosynthesis inhibitor prevents the production of leaf pigments called carotenoids. These are familiar as the colours of autumn when green chlorophyll has degraded. Carotenoids have roles in protecting chlorophyll from being destroyed by more light energy than it can process. Herbicides in this category have several different targets, but all result in treated weeds being bleached white.

For simplicity, a fifth type can be added. This has a single commercial member, glufosinate. Although it is an enzyme inhibitor, it indirectly affects photosynthesis; its faster appearing symptoms are due to membrane destruction resulting in desiccation.

Mode of Action	Effect on Weeds
Photosystem I (PSI) inhibition	• Photosynthesis is affected leading to destruction of cell membranes; the specific effect is much faster than other desiccators. • Leaves quickly yellow and desiccate especially fast in bright sunlight. • Very broad spectrum of weeds controlled; foliar application only; inactivated and immobilised on contact with soil; unique to paraquat and diquat.
Photosystem II (PSII) inhibition	• Photosynthesis is affected leading to destruction of cell membranes, but more slowly than by other desiccators. • Leaves yellow and desiccate from tips, edges and between veins. • Pre- and post-emergence application with soil residual effects; weed spectrum and crop selectivity vary; several different types including: triazines (eg atrazine), ureas (eg isoproturon); nitriles (eg bromoxynil).

Protoporphyrinogen oxygenase (PPO) inhibition	• Rapid desiccation of all green tissue on contact from foliar sprays. • Systemic with slower action when entering through roots. • Biased to broadleaved weed control in various crops, eg fomesafen.
Carotenoid biosynthesis inhibition, eg 4-hydroxyphenyl-pyruvate- dioxygenase (HPPD)	• Leaf pigments cannot be made or protected and so degrade. • Shoots are bleached white. • Pre- and post-emergence application to different target weeds and crops depending on product, eg mesotrione, clomazone, norflurazon.
Glutamine synthetase	• Causes accumulation of ammonia which destroys cell membranes and stops photorespiration and photosynthesis through starvation of amino donors. • Shoots yellow and desiccate, but more slowly than by paraquat. • Controls a broad spectrum of young weeds from foliar sprays only; unique to glufosinate.

MOAs Targeting other Enzymes

In this group, the primary effects are to starve weeds of the essential products normally made by the enzymes affected. The appearance of symptoms is usually slow and depends on how conducive the environment is for plant growth. The ACC-ase inhibitors particularly affect growing points at shoot tips, while the ALS inhibitors and glyphosate have more general effects due to the ubiquity of the amino acids they deplete.

MOA	Effect on Weeds
Acetyl CoA carboxylase (ACC-ase) inhibition	• Stops plants making the fatty acids important in cell membranes. • Growth stops within hours; leaves turn yellow, then purple, brown and die over many days; roots and rhizomes may also be killed. • Post-emergence control of grass weeds in broadleaved crops, eg fluazifop-p-butyl.
Acetolactate synthase (ALS) inhibition	• Stops plants making amino acids leucine, isoleucine, valine, so many proteins can't be made. • Growth stops within hours, stunting shoots and roots; leaves yellow and die over day to weeks. • Control a broad spectrum of weeds in many different crops; post-emergence or pre-emergence, eg sulfonylureas and imidazolinones.
EPSP synthase	• Plants stop making amino acids phenylalanine, tryptophan, tyrosine, so are starved of many proteins. • Growing points in shoots and roots die, leaf chlorophyll degrades, yellowing and exposing purple pigments, stunting; slow in cool climates. • Foliar application only, inactive in soil; unique to glyphosate.

Other MOAs

Generally less is known about the detail of these other MOAs.

MOA	Effect on Weeds
Cell division disruption	• Cells cannot divide for various reasons. • Stunting of emerging seedlings. • Pre-emergence application biased towards grass weed control, eg metolachlor (chloracetamides), pendimethalin (dinitroanilines).

Seedling growth inhibitors	• Various types (eg some prevent the waxy cuticle from forming) but tissues do not develop normally. • Act on new shoots which fail to emerge properly and roots are stunted. • Soil applied for pre-emergence weed control mainly of grass weeds in cereals, eg triallate.
Synthetic auxins	• Hormonal effects on plant growth and development. • Twisting of stems and curling of leaves within hours, yellowing, browning. • Post-emergence control of broadleaved weeds in cereals, eg 2,4-D.

Selectivity

Different plant species respond differently to same herbicide and same plant species respond differently to different herbicides. This is a foundation for phenomenal achievement in modern chemical vegetation (weed) management where objective is to kill weeds and retain others at the same time and place. But selectivity is unwanted within weed species of mixed population. This resulted in buildup of the tolerant species.

The differential response of plants to the herbicide is called selectivity of herbicide. In other wards herbicides harm or kill weeds whereas crop plants are not affected due to selectivity. The fundamental principle is that more toxicant should reach the site of action in active form inside the target plants than in the non target plants. The selective mechanism may occur due to following aspects:

1) Differential rate of absorption of herbicides.
2) Differential rate of translocation of herbicides.
3) Differential rate of deactivation of applied herbicides.
4) Protoplasmic resistance to the specific herbicide

Differential Absorption of Herbicides

In the tolerant wild cucumber (Sicyos angulatus), 2, 4-D absorption was so slow that it kept pace with its metabolism easily than the susceptible cultivated cucumber (Cucumis sativus), thus was tolerant. Similarly, bigleaf maple (Acer macrophyllum) was tolerant to amitrole due to its faster absorption while, bean and lucerne plants were susceptible due to slow absorption of the herbicide by them. Under field conditions, differential absorption of herbicides may occur due to 1) difference in morphology and growth habit of plant species, 2) timing in application of herbicides by different methods, 3) use of antidotes and adsorbents to prevent herbicide absorption by non target plants, and 4) difference in ability of herbicide formulations to contact with non target plants. The selectivity may be due to one or combination of processes.

Differences in Morphology of Plants

Certain morphological features such as narrow upright leaves, corrugated (or) finally ridged leaf surfaces, waxy leaf surface and pubescent leaves allow limited retention of aqueous herbicides on their foliage. Pea, onion, sugarcane, cabbage and colocasia posses the above morphological features. Here the herbicide bounce off as droplets from their foliage or small area may be wetted.

C rops like wheat and sugarcane are protected by herbicide sprays by covered growing point. The limited spray retention provides resistance against selective contact herbicides without any wetting agents.

With translocated herbicides, limited spray retention is not of much help in protecting the non-target plants from herbicide injury. Post emergence application of bromoxynil and ioxinyl controlled many broad leaved weeds in wheat crop due to limited spray retention. Like-wise selectivity of nitrofen in case of rice and Brassica is due to differential wetting. In recent years importance of limited wetting of crop plants as a factor in herbicide selectivity has diminished.

Differences in Growth Habit of Plants

Shoot growth difference: When crop rows have a clear advantage in height over the inter row weeds, directed spraying of herbicides is a common method of achieving selective control of weeds. Herbicide mulches are used in standing crop rows for affecting selective control of germinating weeds.

In slow germinating crops like potato and sugarcane, weeds often establish themselves even before crop emergence; hence they are controlled selectively by spraying a contact herbicide before more than 10% of the crop plants are seen over the ground.

In more advanced stages of crop growth, sometimes specific weed sp may grow much above the crop height. In crops like spinach and Egyptian clover, these tall weeds may often completely hide the crop plants. In the USA. herbicide-laden wax bars have been employed successfully for the control of tall weeds in wide row crops. In the closely sown crops a low volume application of a contact herbicide can be used. In lawns and gardens shoots of nutsedge and other erect weeds can be selectively wiped with herbicides from either with herbicide- laden wax bars (or) clothed stick dipped in concentrated herbicide solution.

Root growth differences: When herbicides are applied to soil, differences in the growth habit of underground parts of weeds and crop plants become important in determining their selective absorption. In general weeds seeds germinate from top 1.25-1.5 cm of soil, whereas many crop seeds are planted 5 to 7.5 cm deep. When a recommended pre-emergence herbicide is applied on the soil surface, and the soil moisture conditions are suitable to leach it to about 2.5 – 3 cm soil depth, it is readily available for absorption to the germinating weeds. Crop plants that grow their roots beyond 5 cm depth obviously avoid herbicide absorption and escape phytotoxicity. This is the basic principle of selectivity of most of the pre-emergence herbicides. Basic principle of selectivity of pre emergence herbicide is a function of herbicide structure, formulation and rate besides soil texture, organisms and inorganic colloids and rain fall.

When any of these factors is unfavorable herbicide may either injure the crops (or) poor weed control is seen. Sometimes, both these adverse effects may occur together when the phenomenon is called 'reverse selectivity'. Selectivity of mollinate between rice and E. colona due to differences in crown root initiation levels of two grasses. CRI is close to surface in E. colona but in rice CRI is below the soils surface in herbicide free zone. Selecitvity of trifluralin in wheat and green foxtail (Setaria viridis) is due to differences in coleoptilar nodes, irrespective of depth from which it is germinated (S. viridis); it is within 1 cm of the soil surface whereas in wheat it is much deeper.

Use of Adsorbents and Antidotes (Induced Selectivity)

- Adsorbents: These are the materials having ability to adsorb herbicides which are placed near crop seed. Activated charcoal is strong absorbent of 2,4-D, EPTC, 2, 4, 5-T, propham, propachlor, pyrazon, trifluralin, chloramben, diuron, butachlor, simazine etc. When a germinating seed is surrounded by a layer of activated charcoal, then seed is prevented from absorption of soil applied pre-emergence herbicides. Mostly in horticultural crops activated charcoal is placed dibble over the crop seeds. Activated charcoal is first used as an adsorbent of 2,4-D. In transplanted horticultural crops, roots of seedlings are dipped in a charcoal before transplanting. Seed pelleting with charcoal has been developed in recent years using gum/ PVA (poly vinyl acetate) for increasing the selectivity of ETPC to maize and cowpea, and of chloramben, butachlor and EPTC to rice.
- Antidote (safener): Safeners are chemicals discovered to antagonize phytotoxicity of specific herbicides to specific plant species. Safeners prove successful against herbicides which inhibit cell division. Otto –L-Hoffman –father of safeners as early as in 1948 observed antagonism of 2,4-D to 2,4,6- T on tomato plants. By 1969, he discovered and reported NA (1, 8 Naphthalic anhydride) as highly successful safener of EPTC and butylate in maize. Effective dose is 0.5g per kg seed. It should be applied as seed dress. Later maize specific safener of EPTC and butylate, namely R-25788 (N, N - dially 1-2, 2, dichloroacetamide) was discovered. The dose of the soil applied R-25788 is 0. 6 kg/ha. It has further been found an antidote of metachlor, alachlor in protecting maize seedlings. A seed coating has been found to provide protection to cultivated oat against pre-emergence alachlor and maize and sorghum against perfluidone and diclofop.CGA-43089 provide safety to sorghum against metalachlor by seed treatment @1-1.5 kg per ha. Use of granules: The granules filters through crop foliage leaving very little for absorption, then settle over ground where the weeds will absorb and has low leach ability. The important desirable character of herbicide granules is low leach ability in soils. Eg. Chlorpropham, Dinoseb, diuron and nitrofen.

Differential Rates of Translocation of Herbicides

Plants can translocate herbicide through the plant as much herbicide it absorbs. When equal amounts of herbicides are absorbed by plants and weeds but translocation rates are different. For example 2,4, 5-T is more toxic to Cucumis trigonus than 2,4-D because it was translocated much more rapidly than the latter compound inside plants. Like-wise differences in the selectivity between sugarcane (tolerant) and beans (susceptible) to 2,4-D was on the basis of its slow translocation in sugarcane and rapid translocation in beans. Always faster translocation does not mean quick killing. In certain cases it will help the plants is escaping specific herbicide action. For instance, diphenamide selective to Convolvulus arvensis because it translocated the herbicide very rapidly from shoots to the roots where it gets metabolized very rapidly than in Avena sativa (it fails to transmit very rapidly from roots to shoots). Soybean has bean found tolerant to oxyflourfen due to its slow absorption.

Differential Rates of Deactivation of Herbicides

Selectivity is primarily a function of differential rate of deactivation. Herbicide selectivity is governed not only by differential absorption & differential translocation but also due to differential rates of deactivation of herbicides by the target and the non target plants. A tolerant plant species deactivates the herbicide molecule rapidly, whereas a susceptible species deactivates it slowly. This deactivation may be a process of i) metabolism ii) Reverse metabolism iii) conjugation.

Reverse metabolism is important mode of herbicide dissipation. Conversion of active herbicide to inactive form is metabolism where as conversion of inactive to active herbicide form is reverse metabolism.

Metabolism

It involves a change in molecular structure of applied herbicides inside the plants yield on phytotoxic compounds. Eg. Ribes nigrum is susceptible to 2,4-D. (It metabolises the 2% of herbicide applied in 96 hours). Whereas Ribes sativum is tolerant to 2,4-D (metabolizes 50% of applied amount within 96 hours). Selectivity of terbacil between Mentha piperata (tolerant) and Ipomea hederaceae (susceptible). Mentha piperata metabolised the herbicide rapidly and shown temporary fall in photosynthesis but in Ipomea hederaceae herbicide persisted for long time to inhibit photosynthesis. Rice is tolerant to bensulfuron due to rapid metabolism inside the plant.

Reverse Metabolism (Inactive to Active)

This is an enzymatic beta oxidation process. Intermediate chemical compounds are more Phytotoxic than original Compounds (parent compounds).Even number carbon ω Phenoxy Alkanoic acid compounds (2,4- DB, MCPB) these are non toxic but in plants they are converted to 2,4-D, MCPA (these are more toxic). This is due to enzymatic oxidation occurs in non-leguminous plants. But in legumes like lucerne, berseem, peas and clovers lack B-oxidation tolerant to 2,4-DB and MPCB.

Conjugation

Coupling of intact herbicide molecule with some plant cell constituents in living plants. Tolerance of grasses and Convolvulus arvensis to 2,4-D, this conjugate with glucose and form glucoside, ß- D glucose ester of 2,4-D. Binding of 2,4-D on certain protein films in tolerant graminaceous members eg.Sugarcane. It takes toxic herbicide concentration out of the main stream and makes tolerant. In Soybean chloramben- translocate rapidly to roots and conjugated with glucose molecules forming N-glucosyl chloramben and an unknown compound Chloroamiben –X.

In apple, maize and certain millets atrazine and simazine are deactivated by conjugation. Enzyme responsible for conjugation in maize is glutathione–S-Transferase. This catalyses conjugation of simazine with reduced glutathione to form S- Glutathion and chloride ion released during this process. Like-wise propanil selective to rice (tolerant); phytotoxic to Echinocloa colonum (susceptible) due to an enzyme called arylacylamine amidohydrolase content in leaves. In Barnyard grass the enzyme is less by 1/60 th as that in rice. In rice, leaves able to hydrolyze propanil to non phytotoxic compounds 3,4-dichloro aniline and propionic acid.

Differential Protoplasmic Resistance

Protoplasm of different plant species differing in withstanding abnormal deficiencies or excess constituents, that may be induced in the presence of some specific herbicide molecules. Eg. Plant show tolerance to dalapon can withstand pantothenic acid deficiency and resist precipitation of cell protein. Buffering mechanism of protoplasm of plants is affected differently by different herbicides. Eg. Tolerance of mustard, groundnut and cotton to trifluralin and nitriles is due to their inherent protoplasmic resistance. Tolerance of rice plants to molinate is due to protoplasmic tolerance.

Multifactor Herbicidal Selectivity in Plants

1. Selectivity of linuron against parsnip in comparison to tomato was due to lower absorption rates and lower pace of metabolism in the parsnip.
2. Selectivity of flurodifen between resistant peanut and susceptible cucumber was found to be due to limited translocation from roots to leaves as well as more rapid metabolism of herbicide that reach the peanut leaves before it could enter the chloroplast. In cucumber flurodifen translocation was fast but its metabolism was slow.
3. Wheat is tolerant to ioxynil and bromaxynil due to limited spray retention, slow translocation and rapid metabolism. Limited spray retention is the first line defense.
4. Distribution of herbicide molecules within the plant is also important factor in the selectivity. Perfluidone and picloram accumulate at the site of action in susceptible plants and equally distributed in tolerant plants. In cotton plant lysigenous glands and trichomes hold high concentrations of triazines and substituted ureas lowering the concentration at the site of action.

Other Selectivity Components

Even if a plant posses some mechanisms to exhibit tolerance to a given herbicide but two important aspects that are to be considered are:

1. Rate of herbicide.
2. Stage of the plant at the time of application in inducing selectivity.

Rate of Herbicides

1. It is important to consider how much and when to apply in obtaining the desired volume of weed control.
2. Under rates improve selectivity to crops at the cost of satisfactory weed control.
3. Over rates decreases the selectivity and cause variable crop injury.
4. In physiologically selective herbicides, range of selective rates is much greater than that is needed for effective weed control. Narrow in case of other herbicides.
5. Most of herbicides loose selectivity at over rates at 0.5-1 kg/ha.
6. Some of herbicides like Dicamba and metamitron loose selectivity even by few grams per ha.

Growth Stage of Crops Plant

There are herbicide susceptible as well as tolerance stages. Except few herbicides which can be applied irrespective of crop stages e.g. simazine in maize and propanil in rice. In most cases at different stages of growth, plants respond differently to a given herbicide. 2, 4-D is selective to winter grains at first 3-6 leaf stage and dough stage. In other stages malformations like onion leafing, missing spikelet ear, rat tail ear occurs. In barley- malting quality is decreased by applying 2, 4-D at susceptible stages.

Herbicide Combinations

Applying two or more herbicides simultaneously, either using prepackage mixtures or by mixing different herbicide products before the application, is a very common approach in intensive agriculture. This is because the application of a single herbicide, even though may provide good control of certain weeds, is often inadequate for satisfactory and cost effective weed control. Furthermore, many herbicides have a narrow spectrum of weed control; whereas, other herbicides do not show the same efficacy against all weeds of their spectrum of control when applied at the recommended rates. Given that weed flora normally consists of many species with varying levels of herbicide sensitivity, more herbicide applications should be often performed or additional measures for weed control should be additionally adopted. This, however, increases the cost of weed control and consequently the cost of crop production.

Mixtures of selected herbicides offer several advantages over the use of a single herbicide, including (a) a reduction in production cost by saving time and labour, (b) a reduction in soil compaction by eliminating multiple field operations, (c) an increase in the spectrum of weeds controlled or an extension of weed control over a longer period of time, (d) an improvement in crop safety by using minimum doses of selected herbicides applied in combination rather than a single high dose of one herbicide, (e) a reduction in crop or soil residues of persistent herbicides by using minimum doses of such herbicides, and selected herbicides.

The use of tank mixtures with two or more herbicide partners presupposes that the combined herbicides behave and act independently (the presence of each one does not affect the activity of the other). In this case, the activity of the applied combination can be easily predicted as the sum of activities of each single herbicide of the mixture when these herbicides are applied separately. In some cases, however, interactions between companion herbicides may significantly modify the biological behaviour of each single herbicide in the mixture. These interactions often result in a reduction or an increase of the activity of the combined herbicides compared with activities when each one of them is applied alone. Practically, the optimum herbicide combinations would be those that exhibit enhanced activity on target weed species and decreased toxicity on crops (increased selectivity). This, however, is difficult to predict since the behaviour of each single herbicide in the mixture is often affected by the presence of the other(s) and the activity of the mixture may also vary considerably depending on plant species, growth stage, and environmental conditions.

Types of Herbicide Interactions

The result of an interaction between two or more herbicides after their application in mixture may be additive, synergistic, or antagonistic. In the first case, the activity of the mixture is equal to the sum of the activities of all herbicides in the mixture when these herbicides are applied separately. In the second and the third case, however, the activity of the mixture is greater or lower, respectively, than the sum of the activities of all herbicides in the mixture when these herbicides are applied separately. It is obvious that in the case of antagonism, where the activity of the mixture is reduced, greater application rates of the affected herbicide are required, whereas in the case of synergism, where the activity of the mixture is enhanced, application rates can be reduced.

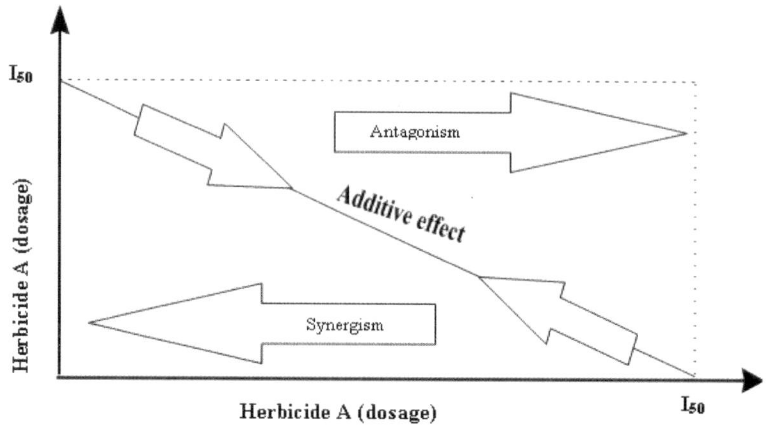

Schematic presentation of herbicide interactions (ID50 = rates of herbicides, applied alone or in mixture, for a 50% weed control).

Antagonistic interactions in herbicide mixtures often cause significant problems in weed control. For example, the application of pyrithiobac in mixture with fluazifop-P has been reported to reduce the efficacy of fluazifop-P on large crabgrass. Similarly, the application of tribenuron in mixture with diclofop has been reported to reduce the efficacy of diclofop on wild oat. It is obvious that such herbicide combinations should be avoided. Antagonistic interactions, however, may be considered beneficial when they reduce herbicide activity on crops. For example, according to Deschamps et al., mixtures of fenoxaprop with MCPA showed reduced toxicity of fenoxaprop on wheat and barley compared with fenoxaprop applied alone. Furthermore, mixtures of thifensulfuron with bentazon showed reduced toxicity of thifensulfuron on soybean compared with thifensulfuron applied alone. Therefore, such herbicide combinations appear desirable unless antagonism on weeds also occurs.

Synergistic interactions may be particularly beneficial when they result in more effective control of troublesome weeds. For example, Flint and Barrett found that mixtures of glyphosate with 2,4-D were more effective on field bindweed (Convolvulus arvensis) compared with separate applications. Similarly, Scott et al. found that mixtures of sethoxydim with dimethenamid were more effective on johnsongrass (Sorghum halepense) compared with separate applications. It is obvious that such herbicide combinations are particularly useful for more effective weed control. Synergistic interactions, however, may cause significant problems when they result in increased herbicide activity on crops. For example, mixtures of ethametsulfuron with haloxyfop, fluazifop, fluazifop-P, quizalofop, and quizalofop-P may cause phytotoxicity and yield losses in Brassica napus and Brassica rapa. Furthermore, mixtures of thifensulfuron (sulfonylurea) with imazethapyr (imidazolinone) may cause phytotoxicity in soybean resistant to sulfonylureas. Mechanisms of herbicide interactions. Interactions in herbicide mixtures can occur prior, during, or after application of the mixture. This means that herbicides may interact physically or chemically in the spray solution or biologically in the plant. Mechanisms of interactions in herbicide mixtures can be broadly grouped into four categories: biochemical, competitive, physiological, and chemical. According to this classification, interactions between herbicides in mixtures may be attributed to a) changes in the amount of an herbicide that reaches its site of action through absorption, translocation or metabolism caused by the presence of the other herbicide, b) interaction at the site of action between the combined herbicides where one herbicide of the mixture affects the binding of the other at its site of action c) interaction between combined herbicides that produces opposite effects on the same physiological process of the plant or synergizes the overall effect, and e) chemical reaction between the combined herbicides that leads to formation of inactive complex or an increase in the rate of metabolism. The aforementioned mechanisms have not been fully documented in many cases. This is because of the great complexity in the study of such interactions since the occurrence of interactions may be a result of two or more mechanisms.

Factors Affecting Herbicide Interactions

The type and the extent of interactions depend primarily on properties of the combined herbicides (chemical group, absorption, translocation, mechanism of action, pathway of metabolism). In general, antagonism has been found to occur three times more often than synergism regardless of the species or the herbicides in which is recorded. Synergism has been found to occur more frequently in mixtures where the companion herbicides belong to the same chemical group. These herbicides normally have similar chemical structure, the same mechanism of action, and similar pathway of metabolism. The high frequency of antagonism in such herbicide mixtures could be attributed in plant inability to metabolize simultaneously two or more herbicides. Antagonism, unlike synergism, has been found to occur more frequently in mixtures where the companion herbicides belong to different chemical groups. These herbicides normally have different chemical structure, different mechanism of action, and different pathway of metabolism. This is because these herbicides probably have a greater chance to interact at the site of action (enzyme or physiological process) or to react chemically and form an inactive complex.

The point of entrance and the mobility of the combined herbicides into the plant may affect significantly the behaviour of the herbicide mixture. In particular, when the combined herbicides enter into the plant through the same point (root or foliage) then the presence of one herbicide in the mixture may reduce the absorbed amount of the other and consequently can reduce its efficacy. Furthermore, the translocated amount of an herbicide to its site of action can be reduced by the presence or the concomitant translocation of another herbicide into the plant. On the contrary, the chance of such interaction is significantly reduced when only one herbicide in the mixture is translocated; whereas, the other is not. A similar trend with aryloxy phenoxy propionate and cyclohexanedione herbicides was observed using data from previous studies; members of both herbicide families were found to be affected more when mixed with systemic rather than contact broadleaf herbicides.

Antagonistic interactions may sometimes be attributed to increased metabolism of an herbicide because of the presence of another herbicide. For example, studies of Jacobson et al. and Shimabukuro et al. showed that the reduced efficacy of diclofop on various species after application with hormone herbicides such as 2,4-D resulted from an increase in its metabolism (formation of complex in the carboxylic group) because of the presence of 2,4-D.

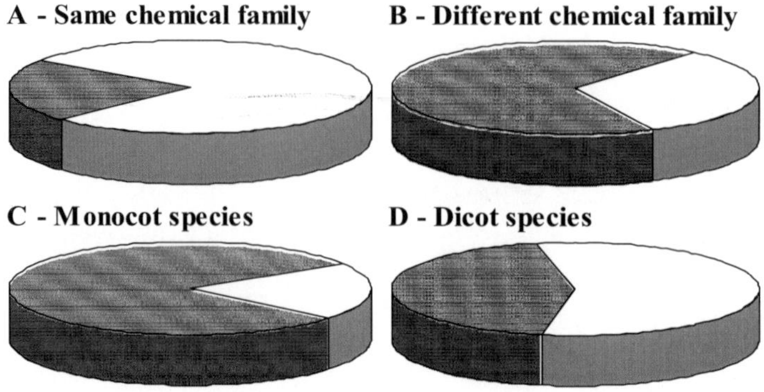

Frequency of antagonistic (z) and synergistic ({) interactions after simultaneous application of herbicides belonging to the same (A) or different (B) chemical groups and applied either on monocotyledons (C) or dicotyledons (D).

The type of interactions between companion herbicides may depend on target plant species. For example, the combination of acifluorfen and bentazon showed an increased efficacy against common lambsquarters and velvetleaf but reduced efficacy against jimsonweed (Datura stramonium) and red root pigweed. Moreover, the combination of herbicides that inhibit acetolactate synthase with herbicides of the diphenyl- ether group showed increased efficacy on prickly sida but reduced efficacy on common cocklebur.

The growth stage of weeds may often affect the extent of interactions between combined herbicides. Liebl and Worsham observed that the post-emergence application of chlorsulfuron and diclofop decreased efficacy of diclofop on italian ryegrass (Lolium multiflorum) and the effect was more severe when the application was performed at the three-leaf growth stage than at the two-leaf growth stage. This may be attributed to reduced detoxification ability from the younger plants and also to their thinner cuticle that probably allowed retention, absorption, and translocation of greater amounts of the applied herbicides.

The post-emergence application of various graminicides in mixture with one or more broadleaf herbicides to broaden spectrum of control often results in reduced efficacy of graminicides. This is the most common case of herbicide interaction reported in the literature and it has been observed in a great number of herbicide combinations on various grass species. Antagonistic interactions between graminicides and broadleaf herbicides are probably due to morphological and physiological differences between grasses and broadleaf weeds. Broadleaf weeds have meristems at the top of the plant; whereas, grasses have them at the base. This difference probably affects absorption and mainly translocation of the foliar applied herbicides particularly the systemic ones that are translocated and accumulated at the meristematic tissues of the plant where they act. Data of the literature show clearly that the simultaneous application of various graminicides with certain broadleaf herbicides limits considerably graminicide absorption by foliage and translocation to the meristematic tissues. This has been confirmed by the results of Zhang et al. who found that the frequency of antagonistic interactions was four times greater than synergistic interactions in grasses; whereas, the corresponding frequencies were almost equal in broadleaf weeds. It is worth mentioning that almost 80%

of the interactions that has been observed in species of the family Poaceae (grasses) refer to cases of antagonism.

It is evident from all the above that many factors may affect the behaviour of herbicide mixtures. Many of the observed interactions are not fully understood at the physiological and biochemical levels and it is possible that more than one mechanisms are involved.

Herbicide Resistant Weeds

The evolution of herbicide resistant weeds is anongoing challenge in modern agriculture. The introduction of herbicide tolerant crops provides opportunities to use different herbicide mechanisms of action to control existing populations of herbicide resistant weeds. However, these new herbicide mechanisms of action may be at risk if farmers do not practice sound integrated weed management practices. This is the same risk confronting the use of all herbicides whether or not crop tolerance is derived from biotechnology, conventional breeding or natural selection mechanisms. This section provides an overview of the origins of resistance, the mechanisms of resistance, and the current status of herbicide resistant weeds globally.

Herbicide resistance is the evolved capacity of a susceptible weed population to withstand a herbicide application and complete its lifecycle when the herbicide is used at normal rates in an agricultural situation. Weed populations may naturally contain herbicide resistant individuals at very low frequencies as a result of rare random genetic mutations. (Herbicides do not cause mutations.) The frequency is dependent on the weed species and the herbicide mode of action. For some herbicides, such as the ALS inhibitors, the frequency of resistant individuals prior to herbicide application may be as high as 1 in 10,000, meaning that ALS inhibitors are prone to a rapid development of resistance. For others, such as EPSP synthase herbicides (e.g., glyphosate), the resistance frequency has been estimated to be less than 1 in a billion. Repeated use of the same herbicide or herbicide mode of action, in the absence of other weed control methods eventually enriches the frequency of these rare mutations to a point where they predominate and cause herbicide failure. Herbicide resistant weeds can then easily spread as contaminants in crop seed and by machinery, water, animals, and wind.

Mechanisms of Resistance

Resistance populations develop from rare random genetic mutations that may exist in individuals within a population. These rare, random genetic mutations provide the weed with a mechanism to resist (tolerate) herbicides. Weeds may resist herbicides through:

- Target site resistance as a result of a modification of the herbicide binding site (usually an enzyme), which prevents a herbicide from binding. If the herbicide cannot bind to the enzyme then it does not inhibit the enzyme and the plant survives. Target site resistance is the most common resistance mechanism. Most but not all cases of resistance to ALS inhibitor, ACCase inhibitor, dinitroanaline, and triazine herbicides are due to modifications to the site of action of the herbicide.
- Enhanced metabolism when the weed develops an enhanced ability to metabolise herbicides into non- toxic or less toxic compounds.

- Decreased translocation which reduces the movement of the herbicide to the site of action.
- Sequestration which stores the herbicide or its toxic metabolites in the cell vacuole, on cell walls, or on tissues remote from the site of action.

From a herbicide resistance management perspective it is important to note that weeds can exhibit cross-resistance and multiple resistance.

- Cross-resistance occurs where a single resistance mechanism confers resistance to several different herbicides classes within a mode of action group. The most common type of cross-resistance is target site cross-resistance, where an altered target site confers resistance to many or all of the herbicides that target that same site.
- Multiple resistance occurs when two or more resistance mechanisms occur within the same plant, often due to sequential selection by different herbicide modes of action, thus resulting in resistance to two or more herbicide modes of action.

Current Status of Resistant Weeds Globally

The "International Survey of Herbicide-Resistant Weeds" recorded approximately nine new cases of herbicide resistant weeds annually, from 1978 to 2009. The tailing off of new cases in the last two years reflects the two years of research needed to confirm resistance, thus some cases investigated in 2007 to 2009 may not be shown in the data.

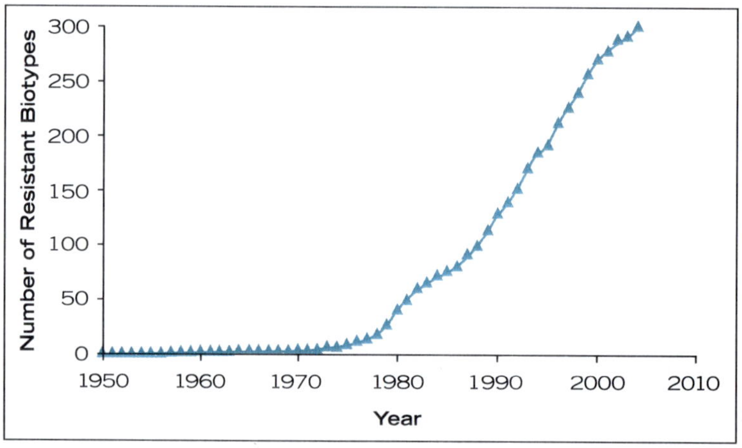

The increase in new cases of herbicide resistant weeds globally.

The importance of resistance cases is based on estimates made by researchers. These estimates are prone to a very wide margin of error, but still give an indication of the number of sites and hectares affected. Many of the 332 cases of herbicide resistance recorded in 59 countries are scientific curiosities rather than major agronomic problems. Of the top 25 most widespread and economically important herbicide resistant weed species ten are grasses, and six are pigweed species (Amaranthus spp.). The most problematic species worldwide is Lolium rigidum, which has been dentified as resistant in 18 countries. It has evolved resistance to nine modes of action, occurs in six ropping regimes (crop and rotation programmes), infests over 9,000 farms and 840,000 hectares. Avena fatua, Amaranthus retroflexus, Chenopodium album, and Setaria viridis are sequentially the next most often reported herbicide resistant weeds globally.

Developing Herbicide Tolerant Crops

Herbicide tolerant crops contain traits that allow them to survive certain herbicides that previously would have injured or destroyed the crop along with the targeted weeds. This allows farmers to use herbicides more effectively and, in some cases, to use less herbicide. Herbicide tolerant crops have been created through conventional breeding techniques or through gene transformation.

Some Important Weed Species

Setaria Viridis

Avena Fatu

Chenopodium Album

Amaranthus Retroflexus

Lolium Rigidum

Biotech-derived Herbicide Tolerant Crops

Biotech-derived herbicide tolerant crops have been genetically engineered through the use of recombinant DNA (rDNA) techniques. Genetic engineering alters the genetic constitution of individual cells by selectively removing, inserting, or modifying individual genes or gene sets using rDNA technology. The term "genetically modified organism" (GMO) is used to describe biotech-derived crops, but does not include herbicide tolerant crops produced through mutations and crossing.

There are two primary methods that have been employed to introduce genetic material into plant cells. They are Agrobacterium-mediated transformation and microparticle bombardment.

Agrobacterium-mediated Transformation

The soil-borne bacterium, Agrobacterium tumefaciens, is able to use genetic engineering processes to transfer parts of its DNA into plant cells. The advantage for the bacterium is that the genetic material it inserts into plant cells causes the plants to produce complex nutrients (opines) which only this bacterium can use as a food source. The inserted DNA also contains plant hormone genes that cause the infected cells to proliferate, resulting in a tumour called crown gall. Research into this phenomenon eventually led to the first published transgenic plant (tobacco) that expressed foreign genes. Since then there have been many additional crops and other plants (maize, tomato, potato, banana, alfalfa, canola, rice, soybean, sugarcane, wheat, etc.) that have been genetically transformed through rDNA technology.

Agrobacterium-mediated transformation of plant tissue is preferred over microparticle bombardment, because it results in higher transformation efficiency, fewer rearrangements, and a low copy number of the transgenes (newly inserted genes). However, the inserts may, in some instances, include bacterial DNA sequences which may complicate the characterization and safety assessments of the biotech-derived events.

Microparticle Bombardment

A useful alternative to Agrobacterium-mediated transformation is microparticle bombardment, a technique used to deliver DNA directly to the host genome. Particles (gold or tungsten) are coated with DNA containing the gene(s) of interest and then fired into plant cells with the hope that a small percentage of DNA dissolves off the particles and becomes integrated into the host genome. This is a less efficient method of producing stably transformed plant cells compared to Agrobacterium transformation, but its advantage is that it can be used on most plant species and only inserts sequences that were on the original DNA segments.

Adjuvants

An adjuvant is any material added to a herbicide spray solution that modifies or enhances the action of that solution. Many herbicides already contain adjuvants as part of their formulations. Some of these formulations can be used directly whereas others need to be applied in conjunction with one or more adjuvants. There are over 3,000 adjuvants available for use. These can be grouped into three general types of adjuvants, including activators, spray modifiers and utility modifiers.

Activators increase the activity of herbicides by modifying certain herbicide characteristics, including particle size of the herbicide spray, distribution of the spray on the plant, spray viscosity, evaporation rate, rate of herbicide uptake or solubility of the herbicide in the spray solution. Activators can be either nonionic (producing little or no ionization in water) or ionic (having a positive or negative charge). It is generally recommended that a cationic (positively charged) herbicide should not be used with an anionic (negatively charged) adjuvant (and vice versa) because oppositely charged compounds could react, diminishing the effects of the herbicide. Most activators have no charge and thus can be mixed readily with any herbicide. There are three categories of activators including surfactants, wetting agents and oils.

Surfactants primarily influence the ability of herbicides to penetrate the leaf's waxy cuticle. Emersed and floating aquatic plants develop waxy cuticles similar to terrestrial plants, whereas submersed plants do not. Most herbicides are prepared in a solution of water. Water is a chemically polar material and thus can be repelled by the waxy surface of leaves. Activators reduce the surface tension of water on plants, and allow the herbicide formulation to wet leaf surfaces and enter into the plant.

Wetting agents increase the ability of water to displace air or liquids from the leaf surface, allowing it to be wet by the herbicide. (Surfactants also have wetting properties but they vary in the degree of wetting they provide.) Wetting agents help spread the solution more evenly over the leaf.

Oils are usually marketed at a concentration of about 80% oil/20% surfactant and are added to water to increase the retention time of a solution on leaves, allowing for an increase in herbicide uptake.

Spray modifiers influence the delivery and placement of spray solution. They confine or alter the characteristics of the spray solution. They include thickening agents (i.e., invert emulsions and polymers), stickers, spreaders, spreader-stickers and foaming agents, which reduce herbicide drift and allow for more exact placement of the herbicide.

Thickening agents modify the viscosity of formulations to reduce or control drift, aid in dispersal and promote sinking. Inverts and polymers are two types of thickening agents commonly used in aquatic herbicide applications.

Invert emulsions are mixtures of inverting oil and water, having a mayonnaise-like appearance on the water surface and a snowflake-like appearance under the water surface. Depending on their solubility, herbicides dissolve in either the oil or water component. The adjuvant/herbicide emulsion sticks to leaves and stems of plants, thus reducing drift and increasing herbicide contact time with plants.

Polymers are long-chain carbon molecules which are up to 40,000 carbons in length, forming a thick mucus- like material which helps to break the surface tension of water and enhance sinking of herbicides. Higher molecular weight polymers are generally formulated as an emulsion and are used as sinking agents. Lower molecular weight polymers are usually formulated as solutions and are used for drift control. Polymers are not very effective in water with a flow rate of greater than 3 cm/sec. as the herbicides/adjuvant mixtures may be washed off leaves before effective contact time is achieved. However, polymers are effective in still waters.

Stickers are made of vegetable gels, resins, mineral oils, vegetable oils, waxes or latex polymers. They promote the sticking of a spray to the sprayed surface. Stickers are usually used for application of fungicides and insecticides rather than herbicides.

Spreaders are blends of primarily nonionic surfactants used for spreading and sticking a spray to plant leaves. They are not as cost-effective as most surfactants but they can increase the effectiveness of some herbicides.

Spreader-stickers are combinations of the above two materials which provide additional retention of herbicide in wet conditions. These adjuvants are more expensive than surfactants and are not used very much in herbicide applications but they are used with fungicides and insecticides.

Foaming Agents are surfactants which are used with specialized spray applicators to create foam for reducing drift and evaporation. These agents are used infrequently for drift control of herbicide applications.

Utility Modifiers are rarely used in aquatic plant control. The addition of modifiers to a herbicide formulation expands the range of conditions (e.g., pH, hardness, etc.) under which a formulation can be used. Types of modifiers include emulsifiers, dispersants and stabilizing agents (including buffering agents and anti-foam agents). Buffering agents and anti-foam agents are used for aquatic plant management. Buffering agents are used to increase the dispersion or solubility of herbicides in alkaline or acid waters used in making up an herbicide solution. Anti-foam agents are mostly silicone-based and are used to eliminate foam in the spray tank, especially useful when mixing herbicides with soft water which usually creates a foaming problem.

Commonly used Adjuvants

Table below contains a partial listing of adjuvants used with aquatic herbicides.

Toxicity

The toxicity of adjuvants is not as well characterized as the toxicity of herbicides. Very commonly, a study of the toxicity of a herbicide focuses on the active ingredient in the herbicide and neglects to consider the toxicity of adjuvants used during application of that herbicide. Part of the reason for the limited toxicity information of adjuvants is that the regulation of adjuvants is not very rigorous.

Name	Type	Action
Big Wet (E,F)	Activator	Nonionic/anion spreader, wetting agent, penetrant
Cide-Kick (E,F,S)	Activator	Nonionic wetting agent, activator, penetrant
Cide-Kick II (E,F,S)	Activator	Nonionic wetting agent, activator, penetrant
Ortho X -77 Spreader (E,F)	Activator	Nonionic spreader, activator
Asgrow "403" Invert Emulsifier (E,F,S)	spray modifier- invert	Invert emulsion, drift control, re duce evaporation, increase droplet spreading and penetration, resist washoff
Bivert (E,F,S)	spray modifier-invert	Invert emulsion, chemical encapsulating, suspending agent, deposition and retention agent, reduce drift and washoff
I'vod Invert ing Oil (E,F,S)	Spray modifier- invert	Invert emulsion, drift control, reduce evaporation, increase droplet spreading and penetration, resist washoff. (Dilution with #2 diesel oil or water required.)
Spra-Mate Invert Emulsion (E,F,S)	Spray modfier-invert emulsion	Invert emulsion, drift control, reduce evaporation, increase droplet spreading and penetration, resist washoff (dilution with #2 diesel oil or xylene required.)
Visko-Rhap (E,F,S)	Spray modifier- inverting oil	Invert emulsion, reduce drift. (Can be diluted with #2 fuel oil or kerosene, if necessary)
Nalquatic (S)	Spray modifier- polymer	Improve sinking, herbicide confinement and contact properties
Nalco-Trol (E)	Spray modifier- polymer	Drift control, developed for rodeo (glyphosate), diquat and 2,4 -d; sinking agent for hydrothal 191 (endothall)
Nalco-Trol II (E,S)	Spray modifier- polymer	Sinking agent developed for hydrothol 191 (endothall) and drift control for rodeo (glyphosate)
Poly Control	Spray modifier- polymer	Drift control, sticking agent, nonionic

Poly Control 2 (S)	Spray modifier- polymer	Drift control, sticking agent, nonionic
Submerge (S)	Spray modifier- polymer	Sinking agent, contact confinement of herbicides (manufactured in both anionic and nonionic forms)

E - emersed plants; S - submersed plants; F - floating plants.

Herbicide Safeners

Herbicide safeners (herbicide antidotes), are chemical agents that increase the tolerance of monocotyledonous cereal plants to herbicides without affecting the weed control effectiveness. The use of safeners offer several benefits to agricultural weed control. Safeners may allow:

(1) The selective chemical control of weeds in botanically related crops.

(2) The use of nonselective herbicides for selective weed control.

(3) The counteraction of residual activity of soil-applied persistent herbicides such as triazines in crop rotation systems.

(4) An increase in the spectrum of herbicides available for weed control in "minor" crops.

(5) An expansion and extension of the uses and marketability of generic herbicides.

(6) The elucidation of sites and mechanism by serving as useful biochemical tools.

The commercial viability of safener concept is indicated by the growing number of herbicide-safener products available on the pesticide market. With the use of safeners, difficult weed control problems can be addressed and without safeners, many herbicidally active substances could have never been applied for weed control.

1,8-naphthalic anhydride (NA) patented by Hoffmann has been considered as the most versatile safener showing less botanical and chemical specificity than other safeners developed later. NA protected cereals as seed treatments against various herbicide chemistries. NA was reported to be mildly phytotoxic to maize (chlorosis and growth inhibition) under some growing conditions. One problem in treating seeds with safeners prior to planting is that phytotoxicity can increase as the time the safener is exposed to the seed increases. With NA, the phytotoxicity to the crop increases with increased time the safener is in contact with the seed during storage. This problem has thus far prevented NA from being introduced to the commercial market.

The introduction of dichloroacetamide derivatives developed as safeners against thiocarbamates and chloroacetanilides was a breakthough in the history of the safeners since these compounds can be applied to the soil in preplant incorporated (PPI) or preemergence (PRE) technology in prepackaged tank mixture with the herbicide. Generally, prepackaged herbicide-safener mixtures offer several advantages over seed safeners. First of all, the manufacturer controls all components of the formulation secondly, the farmers buy and use a single and reliable product which allows a wider selection of crop cultivars. Dichlormid exhibited a remarkable degree of chemical and botanical specificity in protection of maize against thiocarbamates such as EPTC, butylate, vernolate but the safener was less protective to maize against chloroacetanilides.

Table: Structure, logP and application of some important safeners.

Chemical class	Name	logP	Herbicide	Crop	Appl. method
Anhydride	1,8–Naphthalic anhydride (NA)	2.54	Thiocarbamates	Maize	Seed– treatment
Dichloro–acetamide	Dichlormid	1.84	Thiocarbamates Chloroacet-anilides	Maize	PPI, PRE
	Furilazole	2.12	Acetochlor Halosulfuron-methyl	Maize	PRE
	AD–67	2.32	Acetochlor	Maize	PRE
	Benoxacor	2.69	Metolachlor	Maize	PRE
Oxime ether	Cyometrinil	1.56 2.76	Chloroacet–anilides (metolachlor)	Sorghum	Seed– treatment
	Oxabetrinil	2.90 3.64	Chloroacet–anilides (metolachlor)	Sorghum	Seed- treatment
	Fluxofenim		Chloroacet-anilides (metolachlor)	Sorghum	Seed– treatment
Thiazole carboxylic acid	Flurazole		Alachlor	Sorghum	Seed– treatment
Dichloromethyl-ketal	MG–191	1.35	Thiocarbamates Chloroacet-anilides	Maize	PRE
Phenyl–pyrimidine	Fenclorim	4.17	Pretilachlor	Rice	PRE
Urea	Dymron	2.70	Pyributicarb Pretilachlor Pyrazosulfuron–ethyl	Rice	PRE, POST
Piperidine–1–carbothioate	Dimepiperate	4.02	Sulfonylureas	Rice	POST
8–Quinolinoxy–carboxylic esters	Cloquintocet–mexyl	5.03	Clodinafop–propargyl	Cereals	POST
1,2,4–Triazole–carboxylate	Fenchlorazole–ethyl	4.52	Fenoxaprop–ethyl	Cereals	POST
Dihydropyrazole–dicarboxylate	Mefenpyr–diethyl	3.83 3.88	ACCase inhibitors Sulfonylureas	Wheat, Rye, Triticale, Barley	POST
Dihydroisoxazole–carboxylate	Isoxadifen–ethyl		ACCase inhibitors Sulfonylureas	Maize Rice	POST
Arylsulfonyl–benzamide	Cypro–sulfamide	2.09	Isoxaflutole	Maize	PRE, POST

Herbicide-resistant Crops

Herbicide-resistant (HR) crops, particularly glyphosate-resistant (GR) crops, have transformed the way many growers manage weeds. However, after three decades and billions of dollars invested in research, only a few transgenic herbicide traits are commercially available. Two transgenes code for a glyphosate-insensitive 5-enolpyruvylshikimate-3-phosphate synthase (EPSPS; EC 2.5.1.19), the cp4 epsps gene from Agrobacterium tumefaciens strain CP4 and the mutated zm-2mepsps from corn (Zea mays L.), and three transgenes code for metabolic inactivation. One gene from Ochrobactrum anthropi strain LBAA encodes for glyphosate oxidoreductase (GOX), and two homologous genes, pat and bar from Streptomyces viridochromogenes and Streptomyces hygroscopicus, respectively, encode N-acetyltransferases that inactivate glufosinate. Today, HR traits are used on >80% of the estimated 134 million hectares of transgenic crops grown annually in 25 countries with a single trait, CP4 EPSPS, being by far the most utilized.

Growers rapidly adopted the first GR crops because the technology enabled a new weed control practice with glyphosate that was effective, easy-to-use, economical, safe, and novel. The novel attribute of the gene technology was essential to get patents that protected the large investment needed to develop the technology, whereas growers touted the simplicity and convenience of the glyphosate-based crop systems. Initially, glyphosate was exceedingly effective in GR crops, and many growers relied only on glyphosate to control weeds. Some academic weed scientists were concerned about the sustainability of this approach and predicted the evolution of resistance. However, no cases of GR weeds had evolved after more than two decades of broad use in noncrop situations, and some weed scientists and growers began to think that GR weeds would never be a problem. Then the paradigm changed in 1996 with the discovery of GR rigid ryegrass (Lolium rigidum Gaudin) in Australia.

Today, all accept the evolution of GR weeds is threatening the continued success of GR crops and the sustainability of glyphosate. Nineteen weeds have evolved resistance to glyphosate; about half evolved in GR crops. The basis for resistance has been attributed to altered EPSPS target site, reduced translocation or cellular transport to the plastid, sequestration in the vacuole, and gene amplification. GR weeds increase the cost of weed control and diminish the benefits of glyphosate-based weed management systems. In retrospect, it was inevitable that GR weeds would evolve. Glyphosate was a victim of its own success. No matter how effective a herbicide is, weed management programs cannot rely so heavily on one tactic or weeds will ultimately adapt and survive in large numbers.

In essence, GR crops created the "perfect storm" for weeds to evolve resistance. Growers applied glyphosate alone over vast cropping areas to control genetically variable and prolific weeds year after year. Many of these weeds had already evolved resistance to other herbicide modes of action, so there was no good herbicide alternative when these weeds subsequently evolved resistance to glyphosate. Of particular note is the case of the highly competitive and prolific Palmer amaranth (Amaranthus palmeri S. Wats.). The explosion of GR Palmer amaranth populations in the southeastern United States became known as the "pigweed disaster". These GR populations are forcing growers to change their production practices and increase the costs for weed control, even to the extent of hand-weeding. Because of these shortsighted use practices, glyphosate is not as effective as it used to be and growers must supplement glyphosate with other herbicides.

Growers now need to diversify the herbicides they use to mitigate the spread of GR weeds. Unfortunately, the chemical industry has not commercialized a herbicide with a new mode of action (MOA) for over two decades. This is partly because the number of chemicals that must be tested to discover a new herbicide has increased from fewer than 1000 in 1950 to more than 500,000 today and partly because companies are investing less money to discover new herbicides as the widespread use of GR crops has reduced the market opportunity. To address the GR weed problem, the industry is now developing new herbicide resistance traits that will expand the utility of currently available herbicides. However, it is critically important to recognize that these traits represent interim solutions for current weed problems and do not replace the long-term need to discover herbicides with new modes of action and to diversify weed management tactics.

Utilities and Limitations of Current Herbicide Technologies

Current Herbicide use Practices

GR crops came at a time of great socioeconomic change in agriculture. Farm size was increasing, and the number of growers was declining; thus, growers had to become more efficient. Furthermore, weeds were rapidly evolving resistance to various herbicides, and growers perceived weed management as taking too much time. Growers wanted new weed management tactics, and GR crops enabled an economical, efficient, and simple solution. Once growers started using glyphosate, they overused it. The average rate and number of applications of glyphosate increased as its price declined, and the use of other herbicides decreased. Competitors reacted by reducing the price of their herbicides, but those alternatives could not maintain their market presence.

In retrospect, GR crops could have helped to increase the diversity of herbicides that growers used. GR crops did not require that growers use only glyphosate and the added diversity of glyphosate combined with other herbicides would have mitigated the evolution of HR weeds. However, the use of tank mixtures and sequential application of different herbicides declined. In one year, from 1997 to 1998, the use of glyphosate increased 81% in parallel with the increase of GR soybeans from 13 to 36%. The number of herbicide active ingredients used on at least 10% of the U.S. soybean area declined from 11 in 1995 to only 1, glyphosate, in 2002. Even though the chance of weeds evolving resistance to glyphosate in a particular location is still predicted to be lower than that with other herbicides, weeds ultimately did evolve glyphosate resistance as a direct result of the lack of weed management diversification on incredibly large areas of GR crops.

Table: Herbicide types commonly used in corn, soybeans, and cotton and their application method, preemergence (pre) or postemergence (post), with respect to crop.

Herbicide type (group)	Corn	Soybean	Cotton
Glyphosate (G)	PRE and POST	PRE and POST	PRE and POST
Glufosinate (H)	POST	POST	POST
ALS inhibitor (B)	PRE and POST	PRE and POST	PRE and POST
Synthetic auxin (O)	PRE and POST	PRE and POST	PRE and POST
HPPD inhibitor (F2)	PRE and POST	PRE	
PPO inhibitor (E)	PRE and POST	PRE and POST	PRE and POST
ACCase inhibitor (A)		POST	POST
Photosystem inhibitor (C)	PRE and POST	PRE and POST	PRE and POST
Cell division inhibitor (K2)	PRE	PRE	PRE
Phytoene desaturase inhibitor (F3)		PRE	

Interestingly, HR weeds often do not decrease the amount of herbicide used because growers make herbicide decisions based on weed complexes, not individual species or biotypes. If a weed evolves resistance to a herbicide, that herbicide has not lost all of its value as it still controls other weeds, and growers often continue to use the herbicide in a program with another herbicide to control the resistant weed. Furthermore, growers do not "recognize" the potential for weeds to evolve resistance to glyphosate until the biotypes appear in their fields. Unfortunately, this can lead to the practice of sequentially using herbicides until they are no longer effective, which is the fastest way to evolve multiple HR weeds. A combination of herbicides, cultural, and mechanical tactics provides the greatest protection from HR weeds.

Some weed species are particularly troublesome to control and in their propensity to evolve resistance. Problematic weeds in glyphosate-based production systems that have evolved genetic mutations that confer glyphosate resistance include Palmer amaranth and waterhemp. Other weeds such as velvetleaf, morningglories (Ipomoea spp.), Asiatic dayflower (Commelina communis L.), tropical spiderwort (Commelina benghalensis L.), and field bindweed (Convolvulus arvenis L.) often survive because of naturally higher tolerance. Populations of tolerant weed species increase when growers use less than full-labeled rates. Currently, at least seven GR weed species have evolved resistance to multiple herbicide MOAs, with one population of waterhemp in Illinois being resistant to four. The rapid expansion of multiple HR weed populations threatens the sustainability of current crop production systems.

Table: Key row crop weeds and herbicide efficacy.

Weed species		Control rating (0–10) and resistance status						
Common name	Scientific name	Glyphosate	Glufosinate	Als inhibitors	Synthetic auxins	Hppd inhibitors	Ppo inhibitors	Accase inhibitors
Dicotyledons								
Common lambsquarters	Chenopodium album L.	8R	8	7R	9R	9	9	0
Redroot pigweed	Amaranthus retroflexus L.	9	8	9R	9	9	9	0
Waterhemp	Amaranthus tuberculatus(moq.) Sauer	9R	8	9R	8	9	9R	0
Palmer amaranth	Amaranthus palmeri S. Wats.	9R	8	9R	9	9	9	0
Velvetleaf	Abutilon theophrastimedik.	8	8	8–9	8	9	8	0
Common cocklebur	Xanthium strumarium L.	9	9	9R	9	8	8	0
Common ragweed	Ambrosia artemisiifolia L.	8R	9	8R	9	7	9R	0
Giant ragweed	Ambrosia trifidal.	7–8R	8	7–8R	9	8	8	0
Horseweed	Conyza canadensis (L.) Cronq.	7–8R	8	7R	8	8	8	0
Morningglories	Ipomoea spp.	7	8	7	9	7	8	0
Kochia	Kochia scoparia (L.) Schrad.	9R	8	9R	9R	7	8	0

Common sun-flower	Helianthus annu-us L.	9	9	9R	9	9	8	0
Monocotyledons								
Giant foxtail	Setaria faberiher-rm.	9	9	8R	0	8	7	9R
Green foxtail	Setaria viridis (L.) Beauv.	10	8	9R	0	4	5	9R
Yellow foxtail	Setaria pumi-la(poir.) Roemer & J.A. Schultes	9	8	9R	0	6	7	9
Johnsongrass (rhizome)	Sorghum halep-ense (L.) Pers.	9R	6	8R	0	0	8	9R
Shattercane	Sorghum bicolor (L.) Moench	10	9	10R	0	8	7	9R[d]
Large crab-grass	Digitaria sanguina-lis (L.) Scop.	9	8	9R	0	7	6	9R
Barnyardgrass	Echinochloa crus-galli (L.) Beauv.	9	9	9R	0	7	6	9R
Woolly cup-grass	Eriochloa villo-sa (thunb.) Kunth.	9	9	9	0	7	5	8
Fall panicum	Panicum dichoto-miflorummichx.	9	8	8	0	5	4	9R
Italian rye-grass	Lolium multiflo-rum L.	9R	8	8R	0	3	3	9R
Feral corn	Zea mays L.	9R	7R	8R	0	0	6	9R[e]

The best weed management strategy is to control weeds prior to the loss of crop yield potential and proactively delay the evolution of weed resistance. Fortunately, most fields do not have GR weeds yet, and there is still time for many growers to implement diverse and proactive weed management practices. Generally, the basic management tactics are the same for both the prevention and control of HR weeds, that is, diversification of tactics to reduce selection pressure imposed by specific herbicides. The challenge is to implement these practices under prevailing economic constraints when growers are not convinced resistance management tactics will be effective or they believe industry will continue to deliver new solutions to manage weeds. Many growers are reluctant to diversify weed management because they perceive alternative tactics as being less cost-effective despite growing evidence that such tactics can improve profitability as well as mitigate resistant weed issues. More education will help overcome this perception as will the explosion of multiple HR weeds that emphatically persuades growers to diversify their weed management practices now or face serious long-term consequences.

Tactic	Benefits	Risks	Potential impact
Herbicide rotation	Reduced selection pressure, control HR weeds.	Lack of different MOAs, phytotoxicity, cost, limited weed spectrum of alternatives.	Excellent
Herbicide mixtures	Reduced selection pressure, improved control, broader weed spectrum.	Poor activity on HR weed species, increased cost; potential phytotoxicity.	Excellent

Variable application rate and timing	Better control of HR species, more efficient herbicide use.	Lack of herbicide residual activity, timing may be too late to protect yield potential, more applications.	Good to excellent
Adjusted herbicide rates	Better control of target species.	Increased target-site selection pressure with higher rates, increased nontarget site with lower rates (polygenic resistance).	Poor to fair
Precision herbicide application	Decreased herbicide use, reduced selection pressure.	Increased cost of application, unavailability of weed population maps; poor understanding of weed seedbank dynamics; increased variability of control.	Poor
Primary tillage	Decreased selection pressure, consistent efficacy; depletion of weed seedbank.	Increased time required, increased soil erosion, increased costs, additional tactics needed.	Good to excellent
Mechanical weed control strategies	Decreases selection pressure; consistent efficacy, relatively inexpensive.	Increased time required, high level of management skill needed, additional tactics needed, potential for crop injury.	Poor to fair
Crop selection/rotation	Changes agro-ecosystem, allows different herbicide tactics, reduced selection pressure.	Economic risk of alternative rotation crop, lack of adapted rotation crop, rotation crops similar and thus minimal impact on the weed community, herbicides, required, lack of research base, inconsistent impact on HR weed populations.	Fair to good
Adjusted time of planting	Potential improved efficacy on target weeds, reduced selection pressure.	Requires alternative strategies (tillage or herbicide), potential for yield loss, need for increased rotation diversity.	Poor to fair
Adjusted seeding rate	Reduced selection pressure, improved competitive ability for the crop.	Increased seed cost, potentially increased pest problems, increased intraspecific competition, reduced potential yields.	Fair
Planting configuration	Improved competitive ability for the crop, reduced selection pressure.	Unavailability of mechanical strategies, emphasis on herbicides, equipment limitations.	Good
Cover crops, mulches, intercrop systems	Improved competitive ability, reduced selection pressure, improved systems diversity, allelopathy.	Inconsistent effect on HR weeds, lack of understanding about systems, limited research base, potential crop yield loss, need for herbicide to manage the cover crop, lack of good cover crops.	Poor
Seedbank management	Reduced HR weed pressure, reduced selection pressure.	Lack of understanding about seedbank dynamics, requires aggressive tillage, emphasis on late herbicide applications, high level of management skill needed.	Fair to good
Adjustment of nutrient use	Improved competitive ability for the crop, efficient use of nutrient.	Lack of research base, inconsistent results, potential crop yield loss.	Poor

Current Herbicide Technologies

Besides glyphosate, most current herbicides used for weed management in corn, soybean, and cotton are selective and typically used in mixtures to control a broad spectrum of weed species. Here, we provide an overview of the utilities and limitations for various herbicide MOAs that have potential utility in HR crops.

Glyphosate

Glyphosate is a nonselective, broad-spectrum foliar herbicide with no soil residual activity that has been used for >30 years to manage annual, perennial, and biennial herbaceous grass, sedge, and broadleaf weeds as well as unwanted woody brush and trees. Glyphosate is labeled to control over 300 weed species. Many glyphosate formulations and salts are commercially available; the most common salts are the monopotassium and isopropylamine. The type and amount of adjuvant included in the various formulations differ greatly and strongly influence weed control. Glyphosate strongly competes with the substrate phosphoenolpyruvate (PEP) at the EPSPS enzyme-binding site in the chloroplast, resulting in the inhibition of the shikimate pathway. Products of the shikimate pathway include the essential aromatic amino acids tryptophan, tyrosine, and phenylalanine and other important plant metabolic products. The relatively slow MOA and physicochemical characteristics result in glyphosate translocation throughout the plant and accumulation at the vital growing points before phytotoxicity occurs.

Favorable physicochemical characteristics, low cost, tight soil sorption, application flexibility, low mammalian toxicity, and availability of GR crops have helped make glyphosate the most widely used herbicide in the world. A key advantage for glyphosate has been the consistent control of weeds almost without regard to size. However, the flexibility in glyphosate application timing and lack of soil residual have often resulted in growers delaying applications to help ensure that all of the weeds have emerged. Unfortunately, such delay in application means that the weeds have begun to compete with the crop and thus reduced potential yield. The increased use of mixtures with herbicides that have soil residual activity will encourage growers to make earlier glyphosate applications and increase the likelihood that a single application gives season-long control. Other commonly noted weaknesses with glyphosate are higher rates needed to control the more tolerant broadleaf weeds, antagonism by hard water and tank mixture partners, slow speed of action, and poor rainfastness.

Glufosinate

Glufosinate is a nonselective, broad-spectrum foliar herbicide with no soil residual soil activity that inhibits glutamine synthetase, an enzyme that catalyzes the conversion of glutamate plus ammonium to glutamine as part of nitrogen metabolism. Glufosinate is faster acting and controls key broadleaf weeds such as morningglories (Ipomoea spp.), hemp sesbania (Sesbania herbacea (P. Mill.) McVaugh), Pennsylvania smartweed (Polygonum pensylvanicum L.), and yellow nutsedge (Cyperus esculentus L.) better than glyphosate. However, glufosinate is used at higher rates and has historically been more expensive than glyphosate. Cost and more restrictive application timing relative to weed size are probably its greatest disadvantages compared to glyphosate. Because glufosinate behaves as a contact herbicide, it must be applied to smaller plants than glyphosate and is not as effective on perennials that require significant translocation for complete control. Still, glufosinate is labeled to control >120 broadleaf weeds and grasses including key GR weeds. No weeds have been formally reported as glufosinate-resistant yet.

Synthetic Auxins

Synthetic auxin herbicides act as auxin agonists by mimicking the plant growth hormone indole-3-acetic acid (IAA), disrupting growth and development processes, and eventually causing plant death, particularly in broadleaf species. Growers have used auxin herbicides widely for over 60 years as selective herbicides in monocotyledonous crops. Auxins control a broad spectrum of broadleaf weeds, including key weeds that have evolved resistance to glyphosate. Some synthetic auxins such as dicamba have fair soil residual activity with a half-life from 7 to 21 days. Relatively few weeds have evolved resistance to auxin herbicides, which is noteworthy considering their long-term and widespread use. For example, only six weed species have evolved resistance to dicamba after 50 years of widespread use in cereal and noncrop environments.

The increased use of dicamba and other auxin herbicides in auxin-resistant crops has the potential of injuring other broadleaf crops and reducing biodiversity in field edges and nearby noncrop habitat if unmanaged. Off-target movement of auxin herbicides can occur via spray particle and vapor drift. Particle drift is more problematic than vapor drift, but growers can manage with modified application techniques, drift control adjuvants, and correct decisions as to when, where, and how to apply. Particularly troublesome for auxin herbicides would be any movement onto highly sensitive crops such as soybeans, cotton (Gossypium hirsutum L.), or grapes (Vitis vinifera L.). Interestingly, 2,4-D is safer than dicamba on soybeans and dicamba is safer than 2,4-D on cotton. As little as 0.01% of the labeled rate of dicamba can injure soybeans, and 0.001% of the labeled rate of 2,4-D butyl ester formulation can injure tomatoes (Lycopersicon esculentum Mill.) and lettuce (Lactuca sativa L.).

Some forms of dicamba and 2,4-D are highly volatile, especially at high temperatures. For example, the acid form of dicamba is more volatile than amine salt formulations, and some amine salts are more volatile than others. Considerable research is underway to minimize volatilization with new salts and formulations. The manufacturer can also reduce potential off-target movement with application restrictions based on temperature, droplet size, humidity, and wind speed. Because of their volatility and the sensitivity of nontarget crops, growers will probably not use auxin herbicides on vast areas during warm weather as is currently done with glyphosate.

HPPD Inhibitors

The enzyme 4-hydroxyphenyl pyruvate dioxygenase converts 4-hydroxyphenyl pyruvate to homogentisate, a key step in plastoquinone biosynthesis. This is the most recently discovered herbicide MOA, and active analogue testing continues to generate new products. Inhibition of HPPD causes bleaching symptoms on new growth by indirectly inhibiting carotenoid synthesis due to the requirement of plastoquinone as cofactor of phytoene desaturase. Visible injury depends on carotenoid turnover and thus is slower to appear on older tissues than young leaves. HPPD-inhibiting herbicides control a number of important weed species and may have soil residual activity, and no weeds have been formally reported to be resistant to this MOA yet. Corn is naturally tolerant to key HPPD herbicides, but soybeans and cotton are generally sensitive.

ALS Inhibitors

Herbicides that inhibit acetolactate synthase also known as acetohydroxyacid synthase (AHAS), were discovered in the mid-1970s and are still widely used. The ALS enzyme is a key step in the biosynthesis of the essential branched-chain amino acids valine, leucine, and isoleucine. ALS is a nuclear encoded enzyme that moves to the chloroplast via a transit peptide. More than 50 different

ALS-inhibiting herbicides from five different chemical classes (sulfonylureas, imidazolinones, triazolopyrimidines, pyrimidinylthiobenzoates, and sulfonylamino-carbonyl-triazolinones) are commercially available. The characteristics of ALS herbicides vary in their soil residual properties, crop response, and types of weeds that are effectively controlled. ALS herbicides can provide foliar and soil residual activity on important grass and broadleaf weeds at low application rates. The tendency of weeds to evolve resistance to ALS herbicides limits their utility, and their use is now mainly in mixtures with other types of herbicides.

PPO Inhibitors

Protoporphyrinogen oxidase is an essential enzyme that catalyzes the last common step in the biosynthesis of heme and ultimately chlorophyll by the oxidation of protoporphyrinogen IX to protoporphyrin IX. PPO-inhibiting herbicides cause the accumulation of protoporphyrinogen IX, which is photoactive, and exposure to light causes the formation of singlet oxygen and other oxidative chemicals that cause rapid burning and desiccation of leaf tissue. The soil residual and fast action characteristics of PPO herbicides complement the lack of soil residual and the slow activity of glyphosate.

PPO enzyme mutations tend to reduce the enzymatic activity, which helps explain the relatively slow evolution of resistant weeds to this 40-year-old herbicide class. Companies continue to synthesize analogues and commercialize new PPO-inhibiting herbicides. For example, saflufenacil was introduced in 2010 and is labeled for use in a wide variety of crops, including corn, soybeans, and cotton. Its label describes burndown and residual control of 70 broadleaf weeds including key troublesome weeds in glyphosate-based systems such as common lambsquarters (Chenopodium album L.), horseweed (Conyza canadensis (L.) Cronq.), waterhemp, and common (Ambrosia artemisiifolia L.) and giant (Ambrosia trifida L.) ragweeds.

ACCase Inhibitors

Acetyl coenzyme A carboxylase is the first step of fatty acid synthesis and catalyzes the adenosine triphosphate (ATP)-dependent carboxylation of malonyl-CoA to form acetyl-CoA in the cytoplasm, chloroplasts, mitochondria, and peroxisomes of cells. ACCase-inhibiting herbicides generally inhibit the ACCase activity of monocot species and not dicots. The three chemical classes of ACCase inhibitors are cyclohexanediones (DIMs) (e.g., sethoxydim), aryloxyphenoxypropionates (FOPs) (e.g., quizalofop), and phenylpyrazolines (DENs) (e.g., pinoxaden). The ability to use ACCase herbicides selectively in corn would be useful, but the tendency of weeds to evolve resistance to this herbicide class would limit its utility to being part of a weed management system.

Other Herbicide Types

Currently used selective and burndown herbicides will continue to play important roles in weed management in HR crop systems. In addition to the herbicide types already discussed, photosystem II (PSII) inhibitors such as triazine and urea herbicides, lipid synthesis inhibitors such as S-metolachlor, and phytoene desaturase (PDS) inhibitors such as clomazone will continue to be used as crop-selective herbicides to provide soil residual activity on key weeds. Paraquat is a photosystem I (PSI) inhibiting herbicide typically used in conservation and no-tillage production systems for nonselective burndown control of emerged weeds or as a directed spray with specialized application equipment in crop. Paraquat controls a broad spectrum of weeds, and the lack of soil residual allows rotational crop flexibility similar to glyphosate and glufosinate. Paraquat rapidly desiccates leaf tissue and thus does not translocate well enough to control perennial weeds. Paraquat is relatively inexpensive, but its high mammalian toxicity imposes significant use and handling restrictions.

Utilities and Limitations of Current and Future Herbicide-resistant Crop Technologies

Current HR Crop Technologies

A large number of transgenic and nontransgenic HR crops have been commercialized. These HR crops generally eliminated all crop injury concerns and allowed the grower to select new herbicide options with improved weed activity and environmental safety. Before the advent of GR crops, most thought that the utility of HR crops would be limited to complementing selective herbicides. The full impact of HR crops really started in 1996 with the sales of GR soybeans. Since then, the speed at which growers adopted GR crops has been unprecedented in corn, soybeans, and cotton. Success came despite an unpopular "grower contract" and strong objections by biotechnology opponents to potential unknown effects on the environment and human health and the ethical question of interfering with the natural order.

Table: Commercial Herbicide-Resistant Crops in North America.

Herbicide type		Crop	Year available
Bromoxynil		Cotton	1995
		Canola	2000
ACCase inhibitor	- sethoxydim	Corn	1996
	- quizalofop-P	Sorghum	2011
Glufosinate		Canola	1995
		Corn	1997
		Cotton	2004
Glyphosate		Soybean	1996
		Canola	1996
		Cotton	1997
		Corn	1998
		Alfalfa	2005
		Sugar beets	2005
Imidazolinones		Corn	1993
		Canola	1997
		Wheat	2002

		Rice	2002
		Sunflower	2003
Specific Sulfonylureas		Soybean	1994
		Sunflower	2006
		Sorghum	2011
Triazines		Canola	1984

Nontransgenic HR Crops

With the exception of Canada, nontransgenic HR traits are essentially unregulated. Scientists have used a wide range of nontransgenic techniques to create crops with resistance to a number of herbicide MOAs. For example, the first commercial ACCase-resistant crop was a sethoxydim-resistant (SR) corn with an altered ACCase created using tissue culture selection. A second ACCase trait is in the final stages of commercialization for use in sorghum. This trait was transferred with traditional breeding methods from feral sorghum (shattercane, Sorghum bicolor L. Moench) that had evolved ACCase herbicide resistance because of agronomic practices.

Table: Nontransgenic herbicide-resistant crops.

Selection method	Herbicide Type	Crop
Whole plant	Triazine	Canola
Seed mutagenesis	Terbutryne	Wheat
	Sulfonylurea	Soybean
	Imidazolinone	Wheat
		Rice
Tissue culture	Sulfonylurea	Canola
	Atrazine	Soybean
	Imidazolinone	Corn
	Sethoxydim	Corn
Cell selection	Imidazolinone	Sugar beet
Pollen mutagenesis	Imidazolinone	Corn
Microspore selection	Imidazolinone	Canola
Transfer from weedy relative	ALS inhibitor	Sunflower
		Sorghum
	ACCase inhibitor	Sorghum

Creating HR crops for ALS-inhibiting herbicides has been quite successful using tissue culture selection, pollen mutagenesis, microspore selection, seed mutagenesis, and gene transfer from close weedy relatives that had evolved herbicide resistance because of agronomic practices. Today, at least seven different ALS-resistant crops are commercially available. In all cases, resistance is due to an ALS mutation with three general crop phenotypes: broad resistance to ALS herbicides; resistance only to imidazolinone and pyrimidinylthiobenzoate herbicides; and resistance only to sulfonylurea and triazolopyrimidine herbicides.

Glyphosate-resistant Crops

Nontransgenic HR crops were only modestly successful; the big success with HR crops began with transgenic GR soybeans in 1996. Growers perceived glyphosate resistance as the ideal herbicide trait because glyphosate controls over 300 annual and perennial weeds, has flexible application timings, and does not have any rotational crop restrictions. GR crops allowed growers to use glyphosate as an in-crop selective herbicide and replace more expensive, selective herbicides that controlled a narrower weed spectrum and had other issues (e.g., crop tolerance).

Table: Transgenic herbicide-resistant corn, soybeans, and cotton.

Crop	Resistance trait	Trait gene	Trait designation	First sales
Cotton	Glyphosate	*cp4 epsps*	MON1445	1996
		Two *cp4 epsps*	MON88913	2006
		zm-2mepsps	GHB614	2009
	Glufosinate	*bar*	LLCotton25	2005
Corn	Glyphosate	Three modified *zm-2mepsps*	GA21	1998
		Two *cp4 epsps*	NK603	2001
	Glufosinate	*Pat*	T14, T25	1996
Soybean	Glyphosate	*cp4 epsps*	GTS 40-3-2	1996
		cp4 epsps	MON89788	2009
	Glufosinate	*Pat*	A2704-12	2009

Within a decade after glyphosate became commercially available, the search began to find crop resistance to glyphosate. Nontransgenic approaches were not successful, and transgenic approaches were difficult and not initially successful. Initial attempts to find any natural enzymes in crops that could metabolically inactivate or were insensitive at the target site failed. Eventually, a gene for a glyphosate insensitive EPSPS with enzymatic characteristics similar to plant EPSPS was isolated from a common soil bacterium, Agrobacterium tumefaciens strain CP4, which was surviving in a glyphosate manufacturing waste stream in Luling, LA. This cp4 epsps gene has been used to develop GR soybeans, cotton, corn, canola, alfalfa (Medicago sativa L.), bentgrass (Agrostis stolonifera L.), and sugar beet (Beta vulgaris L.).

Glyphosate resistance became the most rapidly adopted technology in the history of agriculture, but the first GR crops were not perfect. The timing, rate, and number of glyphosate applications had to be restricted to ensure crop resistance, and there were reports of a "yield drag". A new generation of herbicide traits currently in development will be combined with current and new glyphosate traits to help continue to improve this technology and extend the transgenic weed management revolution.

Glufosinate-resistant Crops

Glufosinate-resistant crops have been commercially available as long as GR crops but have not been as successful for a number of reasons, particularly because of the higher cost of glufosinate and its more restrictive application timings. Glufosinate resistance is widely available, not only because of its utility as a herbicide trait but also because it has been often used as a marker for other traits, particularly insect resistance traits. Resistance to glufosinate is due to metabolic inactivation of the parent molecule by either of two homologous enzymes, phosphinothricin N-acetyltransferase (PAT) or basta N-acetyltransferase (BAR), that catalyze the acetylation of glufosinate. Both genes were isolated from soil microorganisms, pat from Streptomyces viridochromogenes and bar from Streptomyces hygroscopicus. Cotton and soybean growers who are troubled by difficult to control GR weeds such as Palmer amaranth and waterhemp may rapidly adopt glufosinate-resistant crops and the use of glufosinate. "Dual stack" crop cultivars that include resistance to both glufosinate and glyphosate are now commercially available in cotton, soybeans, and corn and provide growers a choice between two broad-spectrum herbicides as well as an array of naturally selective herbicides to diversify their weed management practices.

Future HR Crop Technologies

Whereas GR crops have been very successful, the evolution of GR weeds was faster and more widespread than many expected. This rapid evolution of GR weeds and the lack of any new selective herbicides with novel MOAs is encouraging HR crop technology to evolve again. The next wave of technologies will combine resistance to glyphosate and other herbicides to provide growers with more herbicide options with different MOAs as well as the possibility of using herbicides with both foliar and soil residual activity. Scientists have discovered a plethora of herbicide traits that can be combined with glyphosate resistance to make multiple HR crops. If used correctly, multiple HR crops with these traits can sustain the usefulness of glyphosate.

Table: Non-glyphosate transgenic herbicide-resistant traits with significant utility in corn, soybeans, and cotton.

Herbicide/Herbicide class	Characteristics
2,4-D	Microbial degradation enzyme.
ALS inhibitors	Resistant ALS from many sources.
ACCase inhibitors and synthetic auxins	Microbial, aryloxyalkanoate dioxygenase.

Dicamba	*Pseudomonas maltophilia*, *O*-demethylase.
HPPD inhibitors	Overexpression, alternate pathway, and pathway flux.
PPO inhibitors	Resistant microbial and *Arabidopsis thaliana* PPO.
Multiple herbicide classes	Glutathione *S*-transferase, *Escherichia coli*.
	P450, *Zea mays*.

Resistance to Synthetic Auxin Herbicides

Corn is relatively tolerant to most synthetic auxin herbicides, but soybeans and cotton are sensitive, and scientists have long sought a transgene to give these crops resistance and allow the use of auxin herbicides. Auxin herbicides control a broad spectrum of broadleaf weeds, including most known GR broadleaf weeds. Because auxin herbicides act rapidly at multiple receptors and compete with an essential plant hormone pathway, making crops resistant by modifying the site of auxin action is difficult. In addition, these receptors respond differently to different auxin herbicide classes, for example, phenoxyacetates, pyridinyloxyacetatesbenzoates, picolinates and quinolinecarboxylates. So far, metabolic inactivation has proven to be a more successful strategy.

A gene encoding for dicamba monooxygenase (DMO), an enzyme that deactivates dicamba, was cloned from a soil bacterium, Stenotrophomonas maltophilla, and used to make dicamba-resistant soybeans. The DMO enzyme encodes a Rieske nonheme monooxygenase that metabolizes dicamba to 3,6-dichlorosalicylic acid (DCSA). The complete bacterial dicamba O-demethylase complex consists of the monooxygenase, a reductase, and a ferredoxin. Electrons are shuttled from reduced nicotinamide adenine dinucleotide (NADH) through the reductase to the ferredoxin and finally to the terminal component DMO. Researchers can successfully express DMO in the cell nucleus with or without a transit peptide as well as in the chloroplasts where the monooxygenase would have a source of electrons produced by photosynthesis and where transgenic proteins can often be expressed at higher levels. Commercialization of dicamba-resistant soybean and cotton is anticipated mid-decade.

A family of aad genes that code for aryloxyalkanoate dioxygenase provides resistance to certain auxin herbicide. The aad-12 gene was isolated from Delftia acidovorans and codes for a 2-keto-glutarate-dependent dioxygenase that inactivates phenoxyacetate auxins and pyridinyloxyacetate auxins (e.g., triclopyr and fluoroxypyr). This trait, DHT2, is being developed in soybeans. A second gene known as aad-1 was isolated from Sphingomonas herbicideovarans and inactivates auxins and ACCase-inhibiting herbicides in the class known as FOPs (e.g., fluazifop). This trait, DHT1, is being developed in corn. Both traits are reported to provide resistance to high rates of 2,4-D with no adverse agronomic effects.

The 2,4-D and dicamba resistance traits will always be used in stacks with at least one other herbicide-resistance trait. The expected increased use of auxin herbicides will increase the potential for off-target movement and injury to sensitive broadleaf plants. Due to this potential environmental problem, the herbicide and trait providers will likely introduce improved herbicide formulations with better use directions before the traits are commercialized mid-decade. Ironically, this risk of off-target movement could drive more rapid adoption of auxin traits because growers will want to protect their soybean and cotton crops from nearby applications of auxin herbicides.

Resistance to HPPD Inhibitors

In some ways, HPPD-inhibiting herbicides are ideal to complement glyphosate. Many HPPD herbicides have soil residual activity and control key broadleaf weeds that have already evolved resistance to glyphosate. Increased resistance mechanisms for HPPD herbicides include a less sensitive target site, overexpression of the enzyme, alternate pathway, increasing flux in the pathway, and metabolic inactivation. Crops resistant to HPPD herbicides have been in field development tests since 1999, but there have been no technical disclosures of HPPD resistance traits under developments thus far. Bayer CropScience in collaboration with Mertec LLC (Adel, IA) and M.S. Technologies LLC (West Point, IA) and Syngenta (Basel, Switzerland) have independently announced plans to develop HPPD-resistant crops. Bayer CropScience recently disclosed that they were developing soybeans resistant to three herbicide types: glyphosate, glufosinate, and HPPD herbicides (e.g., isoxaflutole). Isoxaflutole can provide pre-emergence (PRE) and postemergence (POST) control of a relatively broad spectrum of annual weeds with soil residual activity. The "triple stack" offers the advantage of enabling the use of two herbicide MOAs to which weeds have not yet evolved resistance.

Resistance to other Herbicide Types

Resistance to other herbicide types could also have significant utility. For example, transgenic crops resistant to PPO-inhibiting herbicides have been developed, and the technology even received the trade name Acuron. The first PPO-resistant corn used a double mutant PPO, PPO-1, from A. thaliana. Similarly, PPO-resistant rice used overexpression of the naturally resistant Bacillus subtilis PPO gene to confer resistance. Other approaches including increasing gene copy number and tissue culture to select for overexpression of wild type PPO genes have also been successful. The broad-spectrum weed control and soil residual activity of PPO herbicides could be useful in corn, soybeans, and cotton, but the existing widespread resistance to this class among some Amaranthus species limits the value of the technology.

A transgenic DHT1 trait also gives resistance to ACCase-inhibiting herbicides by degrading the alkanoate side chains to a hydroxyl of the FOP class of ACCase herbicides (e.g., quizalofop). DHT1 corn reportedly tolerates postemergence applications of quizalofop of up to 184 g/ha with no adverse agronomic effects. This trait has utility in corn where commercial ACCase herbicides are not naturally selective. In addition, the specificity of its inactivation could allow the use of other ACCase herbicide types for HR volunteer corn management in rotational crops.

Most herbicide traits only give resistance to herbicides with one MOA. Metabolic inactivation systems based on cytochrome P450 monooxygenases (P450) and glutathione transferase (GST) have the potential to inactivate a wide range of herbicide types. For example, native P450 enzymes can metabolically inactivate acetanilides, bentazon, dicamba, some ALS-inhibiting herbicides, isoxazoles, and urea herbicides. The chemical specificity of this metabolic system may offer the unique potential to allow growers to use herbicides in the same MOA to control weeds in one season and still manage any feral volunteers with a herbicide in the same MOA in the next year.

Multiple HR Crops

No single herbicide resistance trait will be sustainable if the grower uses only the single herbicide type that the trait enables recurrently. The weed problems and their technological resolution must evolve together. Multiple HR crops will help by allowing the use of new herbicide mixtures with multiple modes of action, but agriculture must manage this technology objectively and pragmatically, balancing short-term and long-term interests, so as not to create a "transgenic treadmill".

The lack of soil residual activity has encouraged multiple in-crop applications glyphosate, as many as four or more applications per growing season. Some of the new, multiple HR crop technologies will enable herbicide applications with soil residual activity and thus help growers to reduce selection pressure on the weed community by glyphosate. For example, the glyphosate and ALS trait stack that has recently been deregulated in the United States can allow the use of ALS-inhibiting herbicides with soil residual that are too phytotoxic to use on conventional crop cultivars. This stack consists of a metabolic system to inactivate glyphosate based on an enhanced glyphosate acetyltransferase enzyme from the soil bacterium Bacillus licheniformis (Weigmann) Chester and a highly resistant ALS allele (HRA) with two mutations, tryp574leu and pro197ala.

A wide array of other combinations of current and new herbicide resistance traits is expected within the next decade. If used correctly, these multiple HR crops will provide new uses for existing herbicides to help growers better manage weeds and help sustain the utility of glyphosate and glyphosate resistance traits.

Path Forward

Weed management dramatically changed with the widespread adoption of GR crops. Using glyphosate in GR crops made weed management too simple and convenient. Importantly, the high initial efficacy of glyphosate declined with repeated use, and current glyphosate-based weed management systems are in jeopardy as evidenced by the speed at which weed populations are evolving resistance. Still, glyphosate has not lost all utility; it controls more weeds more effectively than other herbicides, but it can no longer be applied alone anytime on any weed anywhere. Most growers still do not have any GR weeds in their fields and have time to implement proactive HR weed management practices to help sustain glyphosate. However, growers need to act now to diversify the herbicides and tactics they use, the crops they plant, their cultural practices, and field hygiene measures. The flexibility and range of alternative weed management practices will be narrow and require integration to replace glyphosate. These management practices will work better for the prevention rather than the control of GR weeds. Once present, GR weeds can be managed but are difficult if not impossible to eradicate.

Growers need new weed management options now. Current corn, soybean, and cotton cropping systems are based on a heavy reliance on glyphosate. Given the changes in weed populations that are being reported, it is of paramount importance that other weed management alternatives be identified and implemented quickly. It is likely that no new herbicide or trait technology will match the impact of glyphosate and the first GR crops on agriculture. Growers will use these new technologies in combinations to fill in efficacy gaps and diversify weed management practices. Initially, it may look like an attempt to make glyphosate look "as good as it used to be". Some traits such as glufosinate resistance will enable a broad-spectrum alternative to glyphosate. Others will enable options with soil residual activity or new MOAs to control key GR weeds. Some HR crop technologies may benefit from incremental improvements in efficacy and properties of herbicides within long-standing herbicide MOAs that companies are still commercializing.

Growers must diversify their weed management practices now. The more growers diversify, the less the risk that weeds will evolve herbicide resistance. Diversification may make weed management more complex, but growers must not use new HR crop systems in the same way that some used initial GR crops, as a means to rely only on one herbicide until it is no longer effective and then switch herbicides. If growers use the new HR crops and the herbicides that they enable properly, HR crops will expand the utility of currently available herbicides and provide long-term solutions to manage GR weeds.

HR crops will not replace the need for technical innovations, particularly the discovery of herbicides with new MOAs. Diversification will be much easier if growers can chose from among multiple effective and economical weed management options. In areas of the world that have not yet adopted GR crops, growers can learn from the experiences in North and South America. Growers must not wait, but implement best management practices as soon as new trait and herbicide technologies are available. By using diverse weed management practices, growers will preserve the utility of herbicide resistance traits and herbicide technologies and help maintain profitable and environmentally sustainable crop production systems for future generations.

Crop Injury and Herbicides

Crop response, meaning injury, caused by herbicides applied for in-crop weed control can range from no visible response to nearly complete crop loss. Determining the reason or reasons for observed crop injury can be challenging, as several interacting factors may contribute to the severity of response. If the cause is readily discernible, the explanation and prognosis also may be straightforward, but if multiple factors contribute to crop injury, the process of assessment and prognosis may become less precise.

Crop genetics can influence the degree of injury response. Certain corn hybrids, for example, are sensitive to 2,4-D (and other herbicides, for that matter) and may exhibit a great deal of injury following herbicide application. The labels of many corn herbicides, especially postemergence herbicides, have precautionary statements about the potential for certain hybrids to be more sensitive than others to a particular active ingredient. If you are concerned that a particular hybrid may be sensitive to a certain herbicide or herbicide family, contact the seed company representative for information. If more than one formulation of a particular active ingredient is commercially available, the choice of formulation, especially for postemergence applications, also can influence the occurrence of corn injury. For example, ester formulations of 2,4-D tend to be absorbed through the leaf surface faster than amine formulations. Applying 2,4-D esters postemergence with additives such as COC, or tankmixing herbicides with formulations that can "behave" similarly to a spray additive, can increase the rate of 2,4-D uptake into the corn, potentially leading to enhanced corn injury.

The environment has a large influence on the severity of crop injury symptoms from both soil-applied and postemergence herbicides. High air temperatures and relative humidity levels favor enhanced absorption of postemergence herbicides. Adequate soil moisture levels and low relative humidity can enhance uptake of soil-applied herbicides. Rapid herbicide absorption into the crop plant may temporarily overwhelm the plant's ability to break down the herbicide, leading to injury symptoms.

Apart from enhancing herbicide uptake, environment-induced crop stress can enhance crop injury from herbicides. Cool air temperatures and wet soil conditions are good examples of environmental conditions that can induce stress. Why is a crop under stress more likely to be injured by a selective herbicide? In most cases, herbicide selectivity arises from the crop's ability to metabolize (break down) the herbicide to a nonphytotoxic form before it causes much injury. For example, a grass-control herbicide used in corn cannot discriminate between giant foxtail and a corn plant; the herbicide attempts to control the corn just as it attempts to control the giant foxtail. When the corn is growing under favorable conditions, it rapidly metabolizes the herbicide before excessive injury occurs. If, however, the corn plant is under stress (which could be caused by a variety of factors), its ability to metabolize the herbicide may be slowed enough that injury symptoms develop.

The herbicide itself can influence the severity of crop injury, and spray additives applied with a postemergence herbicide or tank-mix combinations may enhance crop response. Always read all label suggestions and precautions related to spray additives that should be either included or avoided when applying herbicides postemergence.

Chapter 4

Effects of Herbicide Use

Herbicides are economical, more efficient and faster than most of the other methods of weed control. However, they can cause harm to crop plants if they are not used properly. They can also have a negative effect on the environment and biodiversity of the area where they are used. All the diverse aspects of the effects of herbicide use have been carefully analyzed in this chapter.

Herbicide Degradation and Decomposition in Soil and Plants

Millions of tonnes of pesticides are applied annually; however, less than 5% of these products are estimated to reach the target organism, with the remainder being deposited on the soil and non-target organisms, as well as moving into the atmosphere and water. The length of time that a herbicide remains active or persists in the soil is extremely important as it relates to the length of time that weed control can be expected. To a certain degree persistence is a desirable property. Herbicides are claimed to persist in the soil as long as their herbicide activity against weeds is wanted. Persistence exceeding this degree is called undue persistence. Also, residual activity is important as it relates to phytotoxic after-effects that may prove injurious to succeeding crops or plantings.

Herbicides are lost from the soil either by physical removal of the unchanged molecule or by degradation. The availability and thus activity, to plants and other organisms, is related to their concentration in the aqueous and/or vapour phases. Seven factors affect the persistence of a herbicide in the soil; Physical removal: adsorption on the soil colloids, leaching, plant uptake and volatility, Degradation: chemical decomposition, micro-organism decomposition, and photodecomposition.

Adsorption of Herbicide by the Soil

Herbicides tend to leave the soil solution and are adsorbed by clay and organic matter particles making the herbicide unavailable for uptake by the weed. Research has shown that:

- Soils high in organic matter require relatively large amounts of pre-emergence and soil sterilant herbicides for weed control.
- Soils high in clay content require more herbicide than sandy soils for pre-emergence or soil sterilant weed control.
- Soils high in organic matter and clay content tend to hold the herbicides for a longer time than sands. The adsorbed herbicide may be released so slowly that the chemical is not effective as a herbicide.

Runoff and Leaching

Runoff moves herbicides in surface water, either mixed in the water or bound to soil particles. The amount of herbicide runoff depends on the grade or slope of the field, the type of soil, the amount of rainfall (especially close to the time of application) and properties of the herbicide. For example, a herbicide applied to a saturated clay soil is highly susceptible to runoff. Established vegetation or plant residues reduce runoff. Herbicide runoff is greatest when heavy rainfall occurs shortly after application. No-tillage, minimum-tillage and soil incorporation reduce runoff. Surface grading, drainage ditches and dikes, and the use of border vegetation can help reduce herbicide movement into surface water.

Leaching is the movement of herbicides through the soil into groundwater. Several factors influence leaching, including water solubility of the herbicide, soil structure and texture, and persistence of herbicide adsorption to soil particles. If a herbicide is strongly adsorbed to soil particles, it is less likely to leach, regardless of its solubility, unless the soil particles themselves move with the water flow. Leaching occurs in any direction (downward, upward, sideways).

The extent to which a herbicide is leached is determined principally by:

- Solubility of the herbicide in water.
- Amount of water passing through the soil.
- Adsorptive relationships between the herbicide and the soil.

In general, those herbicides which are completely water-soluble are most easily leached. Salts of 2,4-D are water-soluble and leach readily through porous, sandy soils whereas esters of 2,4-D are low in solubility and do not leach easily. Herbicides have been known to move upward in the soil. If water evaporates from the soil surface, water may move slowly upward. The water may carry with it soluble herbicides. As the water evaporates, the herbicide is deposited on the soil surface.

Plant Uptake

The uptake of herbicides by plant roots results in their removal from the environment; hence, reduced concentrations in the soil. Absorption is the process by which plants and microorganisms take up chemicals. Once absorbed, most herbicides are degraded within plants. Residues may persist inside the plant or be released back into the environment as the plant decays.

Volatility

Volatilization occurs when a solid or a liquid turns into a gas. A pesticide in a gaseous state can be carried away from the treated area by air currents. This is called vapor drift. Unlike the drift of sprays and dusts that can sometimes be seen during application, vapor drift is invisible. Avoid applying volatile herbicides when conditions favor volatilization, such as temperature inversions. Herbicide labels usually mention the potential for volatility of herbicides. Volatilization can sometimes be reduced through the use of low volatile formulations or soil incorporation of the herbicide.

Herbicides may evaporate and be lost to the atmosphere as volatile gases. The volatile gases may or may not be toxic to plants. The volatile gases may drift to susceptible plants such as those from the ester forms of 2,4-D causing injury to susceptible crops such as tomatoes or grapes. Eptam, Eradicane and Treflan are examples of volatile herbicides which must be incorporated into the soil immediately following application to prevent loss of herbicide to the atmosphere. Rain or irrigation water applied to a dry soil will usually leach the herbicide into the soil, or aid in its adsorption by the soil particles. Once adsorbed by the soil, the loss by volatility is usually reduced. Degradation is, generally, the major route for herbicide dissipation from soil. Herbicide is, gradually, transformed by degradation into series of intermediate metabolites, until the molecule is completely broken down. The intermediate metabolites are particularly important because they can have biological activity, toxicity characteristics and persistence which are different from those of the initial compounds. Degradation is possible in three ways: enzymatically (microbial), photochemical and chemical.

Micro-organism Decomposition

The principal micro-organisms in the soil are algae, fungi, actinomyces, and bacteria. They must have food for energy and growth. Organic compounds of the soil provide this food supply, except for a very small group of organisms that feed on inorganic sources. Microbial degradation occurs when microorganisms such as fungi and bacteria use a herbicide as a food source. Conditions that favor microbial growth include warm temperatures, favorable pH levels, adequate soil moisture, oxygen and fertility. Adsorbed herbicides are more slowly degraded because they are less available to some microorganisms. Micro-organisms use all types of organic matter, including organic herbicides. Some chemicals are easily decomposed (easily utilized by the microorganisms), whereas others resist decomposition.

Chemical Decomposition

Chemical degradation is the breakdown of a herbicide by soil processes not involving a living organism. Adsorption of the herbicides, soil pH, soil temperature and moisture influence the rate of degradation. Some herbicides are more rapidly degraded on low pH soils. Chemical decomposition destroys herbicides through interaction with the soil constituents of oxygen, hydrogen or water. Hydrolysis (interaction with water), for example, is responsible for inactivating atrazine in the soil.

Photodecomposition

Photodegradation is the breakdown of herbicides by the action of sunlight. Herbicides applied to foliage or the soil surface may be broken down by exposure to light. Soil incorporation can reduce herbicide exposure to sunlight. Ultraviolet light from the sun decomposes many herbicides applied to the soil surface. Some herbicides such as fluchloralin, trifluralin are recommended for soil incorporation as they break down readily when exposed to sunlight.

The metabolic fate of pesticides is dependent on abiotic environmental conditions (temperature, moisture and soil pH), microbial community or plant species (or both), pesticide characteristics (hydrophilicity, $pK_{a/b}$, K_{ow}, etc.), and biological and chemical reactions. Abiotic degradation is due to chemical and physical transformations of the pesticide by processes such as photolysis, hydrolysis, oxidation, reduction, and rearrangements. Further, pesticides may be biologically unavailable because of compartmentalization, which occurs as a result of pesticide adsorption to soil and soil colloids without altering the chemical structure of the original molecule. However, enzymatic transformation, which is mainly the result of biotic processes mediated by plants and microorganisms, is by far the major route of detoxification.

Metabolism of pesticides may involve a three-phase process. In Phase I metabolism, the initial properties of a parent compound are transformed through oxidation, reduction, or hydrolysis to generally produce a more water-soluble and usually a less toxic product than the parent. The second phase involves conjugation of a pesticide or pesticide metabolite to a sugar, amino acid, or glutathione, which increases the water solubility and reduces toxicity compared with the parent pesticide. Generally, Phase II metabolites have little or no phytotoxicity and may be stored in cellular organelles. The third phase involves conversion of Phase II metabolites into secondary conjugates, which are also nontoxic. In leafy spurge (Euphorbia esula L.), examples of Phase III metabolism are the conjugation of the N-glycosidemetabolite of picloram with malonate and the formation of a gentibioside from the picloram glucose ester metabolite. There are fundamental similarities and differences between plant and microbial pesticide metabolism.

Table: Three phases of pesticide metabolism.

Initial Characteristics	Properties	Phase I	Phase II	Phase III
Reactions	Parent compound	Oxidation, hydrolysis, reduction	Conjugation	Secondary conjugation or incorporation into bio-polymers
Solubility	Lipophilic	Amphophilic	Hydrophilic	Hydrophilic or insoluble
Phytotoxicity	Toxic	Modified or less toxic	Greatly reduced or non-toxic	Nontoxic
Mobility	Selective	Modified or reduced	Limited or immobile	Immobile
Bioavailability	Readily absorbed in GI tract of animals	Readily absorbed in GI tract of animals	Less absorption	Limited absorption or unavailable

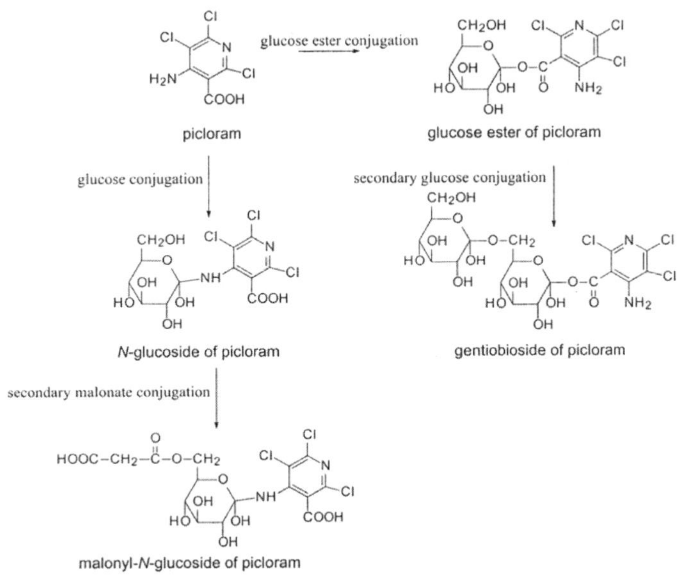

Conjugation and secondary conjugation of picloram in leafy spurge (Euphorbia esula L.).

Herbicide Metabolism

Primary Metabolism

Oxidative Transformations

Oxygenation is the most frequent first step in the bio-transformation of pesticides and other organic xenobiotics. Many of these reactions are mediated by oxidative enzymes, e.g., cytochrome P450s, peroxidases, and polyphenol oxidases. The most extensively studied oxidative enzymes in plants and animals are the P450s, which are the most important enzymes in Phase I pesticide metabolism. Cytochrome P450s are hemethiolate proteins that have been characterized in animals, plants, bacteria, and filamentous fungi. In plants, bacteria, and fungi, P450s produce many secondary metabolites including plant growth regulators, isoprenoids, and alkaloids. Cytochrome P450s are encoded by a superfamily of genes designated as CYP, which have highly conserved residues around the heme portion of the protein. The first plant P450 gene was sequenced in 1990, and presently, more than 500 P450 plant genes have been described. P450 genes occur in clusters in the genome. Regulation and expression of P450s are not well understood in plants or microorganisms mainly because of the very low quantities of P450 enzymes usually present in these cells, particularly if the organism has not been ex- posed to physiochemical, physiological, or xenobiotic stress.

Cytochrome P450s often catalyze monooxygenase reactions, usually resulting in hydroxylation, according to the following reaction: RH 1 O_2 1 NAD(P)H 1 H^1 ROH 1 H^2O 1 $NAD(P)^1$. However, there are many other P450- mediated reactions including dehydration, dimerization, deamination, dehydrogenation, heteroatom dealkylation, epoxidation, reduction, and C–C or C5N cleavage. P450s are divided into three classes. Class I P450s are flavin adenine dinucleotide (FAD) or flavin mononucleotide (FMN) dependent, and reduced nicotinamide adenine dinucleotide phosphate (NADPH) requiring P450s that are usually mi- crosomal membrane-bound proteins in plants and filamentous fungi. Bacteria and nonfilamentous fungi class I P450s are in soluble form. Class II P450s are similar to those in class I, but they are found only in bacterial and animal mitochondria. Class III P450s are located in plant plastids and do not require auxillary redox partners.

Agrochemicals can influence cytochrome P450 systems by acting as effectors, thereby modifying pesticide metabolism, or by modulating overall metabolism of an organism. These effects can increase or decrease physiological activities, which may affect growth and development. Pioneering work on P450-mediated herbicide metabolism in plants was conducted using the phenylurea herbicides, particularly chlortoluron. On the whole-plant level, wheat (Triticum aestivum L.) seedlings exposed to chlortoluron and known cyto- chrome P450 inhibitors (e.g., piperonyl butoxide or 1-aminobenzotriazole) were injured more than plants treated with chlortoluron alone. Similar results were observed using plant cell suspension cultures, where a P450 inhibitor, tetcyclacis, reduced chlortoluron metabolism. Direct evidence that xenobiotic metabolism was mediated by P450s was obtained through experimentation with plant microsomal preparations. Using microsomal preparations from several plant species, it was shown that chlortoluron was metabolized to two metabolites by at least two different P450 enzymes. Since that time, a number of P450-mediated phenylurea- metabolizing genes have been characterized. Mougin et al. demonstrated that the fungicide fenpropimorph was metabolized to an oxygenated metabolite in wheat seedling microsomal preparations. Increased metabolism occurred when seeds were pretreated with naphthalic anhydride, a chemical safener that enhances cytochrome P450 levels. Further, oxidation of fenpropimorph in wheat seedling microsomes was inhibited when the preparations were exposed to carbon monoxide, which binds to the heme portion of the P450 molecule instead of oxygen, thereby blocking enzymatic reactions. These authors sug-gested that fenpropimorph metabolism is P450-mediated. Other researchers have used microsomes to demonstrate that the mechanism of resistance to several dissimilar herbicide chemistries in blackgrass (Alopecurus myosuroides) and rigid ryegrass (Lolium rigidum) was based on enhanced P450-mediated metabolism. Herbicide resistance mediated by P450s may arise via two scenarios: mutation of an existing P450, allowing increased binding and metabolism of the herbicide or increased activity of existing P450s. In the future, researchers will no doubt continue to focus on isolating and characterizing plant P450 genes associated with pesticide metabolism. With a better understanding of P450 genes and their regulation, it may be possible to manipulate the crop plant system to increase herbicide tolerance.

Peroxidases, Phenoloxidases and Related Oxidoreductases

Plants and microorganisms produce a wide range of oxidative enzymes (e.g., peroxidase, polyphenoloxidase, laccase, and tyrosinase) other than P450s that catalyze the polymerization of various anilines and phenols. For example, peroxidase-mediated pesticide transformations in plants that function similar to P450s include decarboxylation, sulfur oxidation, N-demethylation, ring hydroxylation, and aromatic methyl group oxidations. In plants, peroxidase enzymes often function in Phase III metabolism, e.g., formation of bound residues. Horseradish (Amorocia lapathifolia Gilib.) roots contain large quantities of peroxidase. Horseradish root tissue has been used to remove 2,4-dichlorophenol from water and was more effective in contaminant removal than the purified peroxidase enzyme.

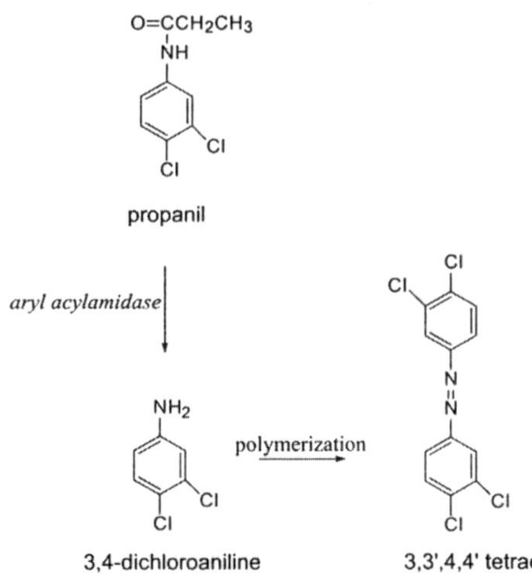

Amide hydrolysis of propanil by aryl acylamidase in plants and micro-organisms and plymerization of 3,4-dichloroaniline to 3,3',4,4'' tetrachloroazobenzene by microbial peroxidases.

White rot fungi (Phanerochaete chrysosporium) offer high potential for xenobiotic transformation becausethey possess free radical–based lignin degrading systems (lignin peroxidase and manganese-dependent peroxidases) that can degrade a wide range of pollutants such as polychlorinated biphenyls (PCBs) and nitroaromatic explosives. In most instances, polymerization products have reduced toxicity compared with the substrate, whereas polymerization of 3,4-dichloroaniline (propanil metabolite) by soil microorganisms results in the formation of carcinogenic tetrachloroazobenzene. Generally, polymerization products are considered to be unextractable humic components.

Oxidative Nitroaromatic Transformations

In microorganisms as opposed to plants, numerous enzymes from many different pathways are capable of oxidizing nitroaromatic compounds, and in many cases the enzymes have been purified, and the genes cloned and sequenced. Oxidative reactions that transform various nitroaromatic compounds from several genera of aerobic bacteria have been described. In bacteria, monooxygenases, flavin monooxygenases, and dioxygenases are generally involved in the initial oxidation of nitroaromatic pesticides, e.g., 2,4-dinitrophenol can be metabolized by these three enzymes. Depending on the compound, nitrite can be released before, or after ring cleavage. The flavin monooxygenase from Sphingomonas strain UG30 is responsible for initial nitrite removal from the herbicide 4,6-dinitrocresol, but not dinoseb (4,6-dinitro-o-sec-butyl- phenol), because of the steric hindrance caused by the bulky butyl group of dinoseb. Overall, these diverse microbial nitroaromatic degradative pathways allow for ring hydroxylation, ring cleavage, and subsequent mineralization of several xenobiotics (e.g., nitrobenzene, nitrobenzoic acid, nitrophenols, and nitrotoluene).

Hydrolytic Transformations

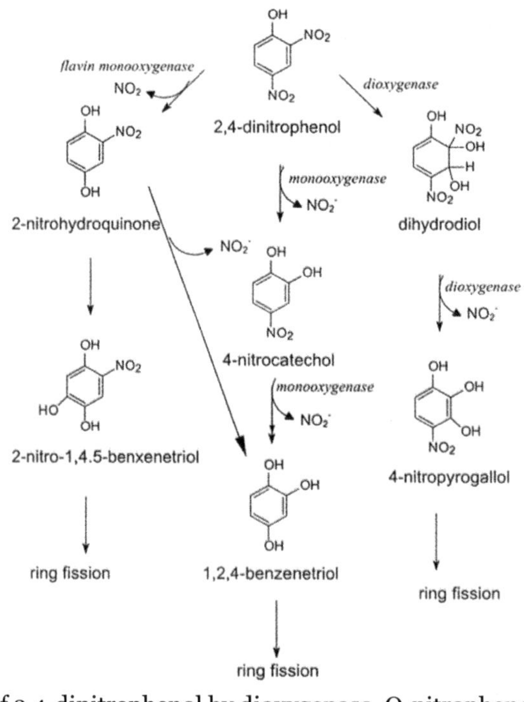

Biotransformation of 2,4-dinitrophenol by dioxygenase, O-nitrophenol-monooxygenase, or pentachlorophenol- flavin monooxygenase characterized in Spingomonas sp. UG30.

Hydrolytic enzymes cleave bonds of a substrate by adding H or OH from H_2O to each product. There are many hydrolytic enzymes that are capable of metabolizing a variety of substrates, particularly those containing amide, carbamate, or ester functional groups. These enzymes may be compartmentalized or extracellular, and reactions strate specificities, thereby allowing degradation of a variety of pesticides. Ester hydrolysis is commonly carried out by esterases and to a much lesser extent by lipases and proteases. Microbial and plant esterases have a characteristic GLY-X-SER-X-GLY motif. The SER acts as a nucleophile, enabling ester bond cleavage. Often, herbicides such as fenoxaprop-ethyl, diclofop-methyl, and 2,4-DB are esterified to increase absorption and selectivity. In plants, the ester bond is metabolized, forming the acid, which is usually more phytotoxic. Depending on the herbicide, de-esterification also can result in immediate herbicide detoxification, as is the case with thifensulfuron-methyl in certain plant species.

Several authors have shown microbial-based ester hydrolysis of diclofop-methyl and fenoxaprop-ethyl in soil by mixed bacterial consortia, and by pure cultures or cell-free extracts. Four types of esterases have been characterized in Pseudomonas fluorescens, each differing in protein structure, cellular localization, and substrate specificity. Although many microbial esterases have been cloned and sequenced, few have been tested for pesticide hydrolysis.

Atrazine was traditionally considered to be moderately persistent in soil; however, in the past several years many bacterial strains representing several genera have been isolated that can completely mineralize atrazine. These authors have suggested that a unique operon of genes encoding for s-triazine degradation has evolved in areas where this herbicide has been used extensively. The gene regions encoding the first three enzymes strain ADP of atrazine degradation have been isolated and characterized from Pseudomonas sp. This bacterium mineralizes high concentrations (500 mg L^{21}) of atrazine under both growth and non-growth conditions, using the herbicide as the sole nitrogen source. The atzA gene encodes atrazine chlorohydrolase, which dechlorinates atrazine hydrolytically to the nonphytotoxic metabolite hydroxyatrazine. The next step in the degradation pathway is hydrolytic removal of the aminoethyl group from hydroxyatrazine by the atzB gene product, hydroxyatrazine ethyl amidohydrolase. Finally, the atzC gene encodes for another amidohydrolase that converts N-isopropylammelide to cyanuric acid. Martinez et al. have recently sequenced the complete catabolic plasmid from strain ADP and have identified three additional genes atzD, atzE, and atzF encoding for cyanuric acid amidohydrolase, biuret hydrolase, and allophanate hydrolase. Thus the total genetic basis for the complete atrazine metabolism in strain ADP has now been identified. Many soil bacteria have the capability to mineralize cyanuric acid. Five other atrazine-degrading bacteria with s-triazine–degrading genes have been identified with. 99% homology to atzABC from Pseudo- monas sp. strain ADP, suggesting that horizontal transfer of atrazine degradation genes may have occurred recently. In fact, in Pseudomonas sp. strain ADP, the three atz genes are on a self-transmissible plasmid pADP-1. Many lines of evidence suggest the ability of microorganisms to mineralize s-triazines developed after the first use of these herbicides in the mid-1950s. In contrast to bacteria, atrazine and other s-triazines are metabolized in plants via N-dealkylation by cytochrome P450s, hydrolytic dehalogenation, or displacement of chlorine with glutathione (GSH). In micro-organisms, there have been no reports of GSH conjugation resulting in dechlorination of s-triazines.

Propanil is the most widely studied pesticide with regard to amide hydrolysis. Rice (Oryza sativa L.) is tolerant to propanil because of high levels of aryl acylamidase, which cleaves the amide bond and is the basis for crop selectivity. Aryl acylamidases are widely distributed in plants, bacteria, fungi, and algae. After 35 year of use, mainly for rice production, propanil-resistant barnyardgrass has developed and is quite widespread throughout many rice-producing regions of the world. Propanil resistance is due to enhanced hydrolysis by aryl acylamidase in resistant barnyardgrass and resistant jungle-rice (Echinochloa colona) biotypes. In several plant species, experiments with propanil analogs (i.e., different ring chloride locations and alkyl chain length, etc.) revealed that generally propanil was the preferred aryl acylamidase substrate. Synergistic interactions can occur when propanil is mixed with any of several agrochemicals (e.g., carbamate and organophosphorus insecticides, and the herbicides anilofos, pendimethalin, and piperophos). Synergism of propanil and the insecticide carbaryl was the result of competitive inhibition with aryl acylamidases; however, the mecha- nisms of propanil synergism with other agrochemicals have not been fully characterized. Likewise, certain agrochemicals, e.g., carbamates, can inhibit the hydrolysis of propanil in soil and water by inhibiting microbial aryl acylamidase activity.

Table: Comparison between plant and microbial pesticide metabolism.

Biotransformation	Plants	Microorganisms
General pesticide metabolism	Detoxification	Mineralization
Oxidation	P-450 mediated	Not generally P-450 mediated, Mediated by various oxidoreductases
P-450 oxidation	Microsomal membrane bound	Soluble form, not membrane bound
Hydrolytic transformation	Predominantly via esterases, amidases, aryl acylamidases, and nitrilases	Greater enzyme diversity
C–P bond cleavage	None known	Diverse C–P lyases and hydrolytic enzymes
Aromatic nitro-reductive processes	Nitroreductases GSH conjugation	Nitroreductases, No GSH conjugation
Reductive dehalogenation	None known	Halo-respiration
Conjugation	With sugar and amino acids Compartmentalized or sequestered GSH conjugation	With xylose, methyl, or acetyl groups Conjugates formed extracellularly, No known GSH conjugation.

The substrate specificity of microbial aryl acylamidases varies even more considerably than that of plant aryl acy- lamidases. For example, in some P. fluorescens strains, the substrate range is limited to the acylanilide pesticides, but Bacillus sphaericus has a wide substrate range, including acylanilide, phenylcarba-mate, and substituted phenylurea pesticides. Many microbes are capable of amide hydrolysis of propanil. In one study, 37% of 97 bacterial isolates collected from soils and rice flood water of the Mississippi Delta (an area where propanil has been widely used) were capable of hydrolyzing propanil. Aryl acylamidases have been purified from several bacterial genera including B. sphaericus, P. fluorescens, Pseudomonas pickettii, Pseudomonas aeuruginosa, Nocardia globerula, and a coryneform-like bacterium. These enzymes range in size from 52.5 to 127 kDa and differ with respect to the subunit aggregation, i.e., some are monomers, dimers, or tetramers. All amidase proteins have a characteristic hydrophobic GLY- GLY-SER-SER motif.

Organophosphorus pesticides are hydrolyzed by micro-organisms and have been extensively studied in Pseudomonas diminuta and Flavobacterium sp. hydrolysis, oxidation, and glutathione biotransformations of organophosphorus pesticides appear to be equally important detoxification mechanisms in plants. In plants and bacteria, there is limited literature on the role of phosphatases and sulfatases in pesticide metabolism. However, there is evidence that sulfatases in the fungi Trichoderma harzianum and P. chrysosporium hydrolyze the insecticide endosulfan. Nitrile hydrolysis is the main route of metabolism of bromoxynil in wheat and of cyanazine in wheat and potato. Hydrolysis of the nitrile group produces an amide moiety that is converted to carboxylic acid, which may be subsequently decarboxylated. In contrast to plants, several bacteria species hydroxylate the cyano group of bromoxynil. Hydrolysis of the carbamate moiety of phenylcarbamate pesticides is common in animals and soil microorganisms but not in plants. In plants, the major metabolic route for the phenylcarbamate pesticides CIPC and IPC is aryl hydroxylation and conjugation, rather than hydrolysis of the carbamate moiety.

Generally, there is more known about xenobiotic hydrolysis in microorganisms than in plants. However, the precise physiological role of many hydrolytic enzymes is not known. There is a need to further understand the mechanism and regulation of hydrolytic enzymes.

Aromatic Nitroreductive Processes

Generally, nitroaromatic compounds are transformed differently in plants in comparison with microorganisms. For example, the major metabolite of trifluralin in peanut (Arachis hypogaea L.) is N-depropylated trifluralin, whereas in sweet potato (Ipomoea batatas L.), the monoamino-derivative of trifluralin is predominant. In contrast, trifluralin is transformed via nitroreductase by microbes. In plants, glutathione conjugation of pentachloronitrobenzene occurs concomitant with the removal of Cl or NO_2. Although glutathione-mediated displacement of the nitro group of aromatic compounds has been described in plants, it has not been reported in microorganisms.

In bacteria, three pathways of reductive metabolism of nitroaromatics have been characterized: aromatic nitroreduction, partial nitroreduction, and hydrogenation. Reductive metabolism of nitroaromatic xenobiotics is mediated by nitroreductase enzymes found in aerobic and anaerobic bacteria, and several genera of fungi. Nitroreductases are flavoproteins that use NAD(P)H as reducing equivalents, require FMN/FAD as cofactors, and have varying sensitivities to O^2 con- centrations. Some bacteria contain multiple aromatic nitro-reductase isozymes. It is sometimes difficult to separate biological and chemical xenobiotic reductions because reduction of aromatic nitrogroups, e.g., trifluralin and diphenyl ether herbicides, may be coupled with anaerobic reduction of humic acids or iron reduction. The conversion of the herbicide acifluorfen to aminoacifluorfen is a common example of an aromatic nitroreduction reaction catalyzed by bacteria under aerobic and anaerobic conditions, as well as in Enterobacter cloacae and P. fluorescens cell-free extracts. Aminoacifluorfen is susceptible to sorption and incorporation into soil humic material. There is potential to develop transgenic crops that express a bacterial nitroreductase gene to metabolize diphenyl ether herbicides, thereby providing crop tolerance to these herbicides.

Numerous bacteria are capable of partial nitroreduction, resulting in NH_3 release and subsequent ring cleavage. Partial nitroreduction pathways are catalyzed by a nitroreductase that reduces the nitro moiety to a hydroxylamino group, followed by further molecular rearrangement catalyzed by a hydroxylaminolyase, forming the hydroxyl amino derivative. Although bacterial partial nitroreduction of several xenobiotics including p-nitrobenzoate and nitrobenzene has been demonstrated, partial nitroreduc-tion of pesticides has not been reported. In bacterial partial reductive hydrogenation reactions, the nitroaromatic compound is used as the sole carbon or nitrogen source.

Carbon–phosphorus Bond Cleavage Reactions

Organophosphonates used as pesticides, antibiotics, lubricants, and flame retardants have a carbon-to- phosphorus (C–P) bond, which does not undergo photochemical, hydrolytic, thermal, or chemical degradation. However, many organophosphonate compounds do not persist in the environment because of microbial degradation. Currently, it is believed that plants do not possess the ability to break the C–P bond of organo-phosphonates, and relatively little is known about fungal organophosphonate metabolism. However, degradation of C–P bonds has been extensively studied in bacteria. For instance, a gene cluster designated phn, consisting of 17 genes from Escherichia coli, is responsible for the degradation of a wide range of phosphonates and is likely to encode for a C–P lyase. The enzyme(s) responsible for direct cleavage of organophosphonate C–P bonds is known by the general name C–P lyase. The ability of C–P lyase to degrade a wide variety of chemically diverse phosphonates is quite striking. However, the precise mechanism of C–P lyase is not fully understood. It is hypothesized that alkane-phosphonate biodegradation occurs by two different pathways in which either organophosphonyl or organo-phosphoranyl radicals are formed.

With regard to herbicides, the two-carbon phosphorus bond (C–P–C) of glufosinate is difficult to cleave, and although glufosinate is metabolized in soils, it is not known if the C–P–C bond is broken. In contrast, many researchers have reported the microbial mineralization of glyphosate in the by gram-negative and gram-positive bacteria, under both anaerobic and aerobic conditions and with no lag phase of degradation. Numerous bacterial strains can use glyphosate as the sole P source without mineralizing it. However, only an Achromobacter strain and a Streptomyces sp. were able to use glyphosate as the sole carbon or nitrogen source via C–P bond cleavage and formation of sarcosine constitutively in pure culture. This research indicates that a consortium of microbial species may be required for glyphosate mineralization or that gly- phosate is metabolized by fastidious bacteria. Two main pathways of glyphosate C–P bond cleavage have been characterized; however, neither reaction has been solely used for generating commercially viable genetically engineered glyphosate-tolerant crops. In one pathway, initial cleavage of the C–P bond yields inorganic phosphorus and sarcosine, and the latter is further converted to glycine and a C1-unit. In the second case, glyphosate oxidoreductase (GOX), a well-characterized 46.1-kDa flavoprotein, cleaves glyphosate into glyoxylate and aminomethylphosphonic acid. Aminomethylphosphonic acid is further degraded by a C–P lyase.

Pesticide Conjugation Reactions

Carbohydrate and Amino Acid Conjugation

Hall et al. recently defined pesticide conjugation as the "metabolic process whereby an exogenous or endogenous natural compound is joined to a pesticide or its metabolite(s) facilitating detoxification, compartmentalization, sequestration, and/or mineralization."Conjugation of pesticides often involves utilization of existing enzymatic machinery and is therefore called a cometabolic process. Glucose conjugation to pesticides occurs primarily in plants, resulting in several metabolites including O-, S-, and N- glucosides, glucose ester, gentibioside, and malonyl-glucose conjugates. The most common glucose conjugates are O-glucosides because pesticide oxidation reactions form hydoxyl groups, which are suitable sites for glucose conjugation.

Differential conjugation of 2,4-D imparts differences of susceptibility in wheat and some broadleaf species. Many susceptible broadleaf weeds produce glucose ester metabolites, which are readily susceptible to hydrolysis, yielding phytotoxic 2,4-D. Conversely, 2,4-D–tolerant wheat rapidly produces amino acid conjugates and O-glucosides, which are stable nonphytotoxic metabolites that are not easily hydrolyzed. Amino acid conjugation occurs primarily in plants and is very common. Most of the research on amino acid conjugation of pesticides has been conducted on 2,4- D. Twenty amino acids have been found to conjugate with 2,4-D.

Uridine diphosphate–glucosyl (UDPG) transferase, an enzyme involved in cellulose biosynthesis, mediates pesticide–glucose conjugation and pesticide–glucose ester conjugation reactions. As mentioned above, glucose esters of pesticides are cleaved by esterases, often resulting in the release of the pesticide. However, the addition of a second glucose molecule to the glucose ester produces a gentiobiose conjugate, which is not readily hydrolyzed. Other complex sugar conjugates in addition to gentibioside (two glucose molecules) are glycosides (a glucose and one other sugar, such as arabinose).

Pesticide–sugar conjugates can undergo further conjugation with malonate via reaction with malonyl CoA, a common reaction in higher plants. In tomato, the herbicide metribuzin is conjugated to glucose, which is subsequently conjugated to malonate, forming the N-malonyl–glucose conjugate. A range of UDPG transferase activity within various tomato cultivars confers differential tolerance of these cultivars to metribuzin. Furthermore, increased metribuzin phytotoxicity in all the cultivars was noted under low light conditions. It was speculated that under low light conditions less glucose and UDPG were produced, thereby reducing conjugation and elevating herbicide phytotoxicity.

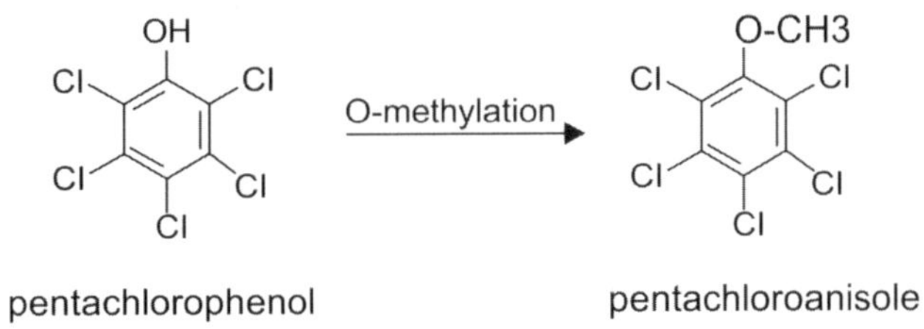

Conjugation of pentachlorophenol by microorganisms.

Microbial Pesticide Conjugation

Microbial pesticide conjugation reactions include xylosylation, alkylation, acylation, and nitrosation and can occur intra- or extracellularly. During fungal degradation of lignin, carbohydrates are generated, but toxic phenols are also concomitantly released. These phenols are extracellularly conjugate to xylose as a detoxification mechanism. Fungi use the same process to extracellularly conjugate 2,4-D and 2,4,5-T with xylose.

Fungi generally biotransform pesticides and other organic xenobiotics by inducing minor structural changes to the pesticide, rendering it nontoxic. The biotransformed pesticide is released into the soil, where it is susceptible to further metabolism by bacteria. Both fungi and bacteria use methylation as a conjugation reaction to detoxify xenobiotics. For example, formation of O-methylated pentachlorophenol by fungal cultures of Trichoderma virgatum and many gram-positive and gram-negative bacteria results in a less toxic, but more recalcitrant, pentachloroanisole. Phanerochaete chrysosporium methylates chlorophenoxyacetic acid via a manganese–lignin peroxidase, which is an extracellular degrading enzyme system.

Pesticide conjugation via microbial acylation with acetate or formate is also common. Phenols and anilines in soil, which are typical breakdown products of phenylacylanilides, phenylcarbamates, and substituted phenylurea pesticides, are often acylated by fungi. For example, the herbicide meto-bromuron is metabolized by microbes to 4-bromoaniline, which is metabolized to 4-bromo-acetanilide. Nitrosation is a process mediated by bacteria, wherein nitrite reacts with a secondary amine to form a nitrosamine derivative. Nitrosamines can be generated by both enzymatic and nonenzymatic processes. Certain pesticides are converted to secondary amines, e.g., dimethyl and dimethylamines, when metabolized in soil.

Plant Glutathione Conjugation Reactions

Glutathione commonly present in the reduced form, is ubiquitously distributed in most aerobic organisms. Homoglutathione a GSH analog, occurs in several legume species. Although GSH concentrations vary during plant development, GSH is found in relatively high concentrations in most plant tissues. Glutathione is phloem mobile and is degraded by carboxypeptidases and transpeptidases in the cytoplasm and vacuoles. Generally, GSH synthesis is limited by availability of cysteine and hence by the concentration of sulfate ions.

Nonenzymatic GSH conjugation may be important for the metabolism of several herbicides. For example, increased GSH concentrations protected wheat from fenoxaprop injury. This reaction was considered nonenzymatic because glutathione S-transferase (GST) activity in these plants was low. However, enzymatic conjugation of xenobiotics with GSH via GSTs is more common than nonenzymatic conjugation.

Glutathione-S-transferases are homo- or heterodimer, multifunctional enzymes located in the cytosol, which catalyze the nucleophilic attack of the sulfur atom of GSH by the electrophilic center of the substrate. More than 50 plant GST gene sequences from 13 plant species have been published. Com- pared with other plant and bacterial species, corn GST gene enzyme systems have been the most exten- sively studied. X-ray crystallography revealed that the N-terminus of this dimeric enzyme is highly conserved and binds GSH at the G-site. The less conserved C-terminal is an a-helix that binds substrates, including herbicides, at the H-site. These two binding domains are kinetically in- dependent. Recently, a new phylogenetic plant GST classification system was proposed by Dixon et al. and Droog that consists of four classes.

In plants and animals, regulation mechanisms and the catalytic function of GST enzymes have been highly conserved during evolution. Some GSTs are constitutively expressed in certain tissues, but GST regulation can be modified by agrochemicals, including herbicide safeners and synergists. It is hypothesized that plant GST gene promoters have multiple regulatory elements that respond differently to specific or more general stress-related signals. Class I corn GSTs have safener responsive elements, designated by an ATTTCAAA nucleotide sequence. Moreover, GSTs probably have common mechanisms of signal transduction to activate gene expression; e.g., all active oxygen species may affect a common transduction pathway during oxidative stress.

The role of GSTs and GSH in plants encompasses several major functions. The first is the metabolism of secondary products, including cinnamic acid and anthocyanins. A second function is regulation and transport of both endogenous and exogenous compounds, which are often GS-X tagged for compartmentalization in the vacuole or cell wall. This is a particularly important aspect for herbicides, anthocyanins, and indole-3-acetic acid. Protection against oxidative stress from herbicides, air pollutants, pathogen attack, and heavy metal exposure is a third function. Glutathione conjugates and their terminal metabolites are stored in the vacuole or bound to the cell wall. Glutathione conjugate pumps in the tonoplast membrane carry GSH conjugates across the membrane. In the vacuole, peptidases release the glutathionyl moiety.

Glutathione-S-transferases in plants were first studied be-cause of their ability to detoxify herbicides. Glutathione-S-transferase–based herbicide metabolism imparts herbicide selectivity in several plant species. Many herbicide families, in- cluding sulfonylureas, aryloxyphenoxypropionates, triazinone sulfoxides, and thiocarbamates, are susceptible to GSH conjugation. Furthermore, there is a positive correlation of both GSH levels and the activity of spe-cific GST enzymes with the rate of herbicide conjugation and detoxification. For example, the resistance of a velvetleaf (Abutilon theophrasti Medicus) biotype to atrazine was the result of an enhanced rate of GSH conjugation. To study GST substrate specificity, Sommer and Boger purified four recombinant N-terminal 63His- tagged corn GST isoforms, using an E. coli expression system. The recombinant GST isoforms included GST types I, II, III, and IV, with subunits of 29/29, 27/29, 26/26, and 27/27 kDa, respectively. Substrate specificity for each of the four isoforms was different and was based on subunit specificity. For example, GST isoforms with a GST29 subunit could readily conjugate 1-chloro-2,4-dinitrobenzene, whereas isoforms with GST27 subunits had their greatest metabolic activity on thiadiazoliodine and metazachlor. Moreover, the GST27 subunit had metabolic activity on endogenous hy-droperoxides such as linolenic acid and cumene hydroper-oxide. These results suggest that certain GSTs function through peroxidase activity, to protect the plant from oxidative stress. Based on metabolism levels reported in the literature, the 63His-tag expression in E. coli does not seem to affect GST isoform substrate specificity and is therefore a convenient system to study GST-mediated herbicide metabolism.

Bacterial Glutathione Conjugation Reactions

Compared with plant GSTs, few bacterial GSTs have been characterized at the biochemical level. It is thought that bacterial GST–mediated herbicide metabolism is important because herbicide metabolites with thiol, thioester, and sulfoxide moieties have been identified in soil. The role of bacterial glutathione conjugation has been demonstrated in the dechlorination of chloroacetamide herbicides, e.g., alachlor and metolachlor, and the ether bond cleavage of the herbicide fenoxaprop-ethyl. Glu- tathione-S- transferases that function as reductive dehalogenases from Sphingomonas strains are involved in the dechlorination of pentachlorophenol and lindane. In spite of a few well- characterized degradation schemes, little is known about bacterial GST regulation and function. Most of the knowledge about bacterial GSTs is based on genomic analysis. Many gene sequences with homology to corn GSTs have been identified within bacterial genomes; however, there is a need to discriminate and determine enzyme function at the biochemical level.

Two messages become clear from bacterial genomic research there is a large set of GST- homologous genes, which vary in size and content in bacteria and certain GST-classified genes are associated with operons and gene clusters involved in xenobiotic dehalogenation. In E. coli and P. aeruginosa genomes, there are 8 and 17 GST-like genes, respectively. In both organisms, only four of the GST-like genes have 40% homology with known plant and mammalian GSTs at the protein level. The P. aeruginosa genes, however, have greater sequence similarity to known biodegradation GST genes, even though the physiological roles of these genes are unknown. Nevertheless, known bacterial GSTs have structural similarities to plant and mammalian GSTs despite the extensive variation in sequences. The bacterial GST dehalogenases thus far identified include dichloromethane dehalogenase, tetrachlorohydroquinone reductase involved in pentachlorophenol metabolism, and 2,5-dichlorohydroquinone reductive dehalogenase involved in lindane degradation. An unusual function of a GST enzyme from a Rhodococcus strain is the ability to open the epoxide ring during the degradation of isoprene and chlorinated ethenes. This GST enzyme was purified and the iso1 gene cloned and characterized. Similarly, a human GST has been characterized that acts as both an isomerase and a dioxygenase in aromatic ring opening. This suggests that bacterial GST dehalogenases also may act as isomerases and dioxygenases in aromatic ring. There are many unanswered questions regarding bacterial GST–mediated xenobiotic metabolism, GST regulation, as well as GSH–conjugate uptake, excretion, and toxicity. Moreover, there is potential for using bacterial GSTs in bioremediation and biodetoxification; however, further research is required to fully understand the function and substrate specificity of bacterial GSTs.

Formation of Bound Pesticide Residues

Pesticides (mainly conjugated pesticides) are often bound to plant cell walls. Bound pesticide residues are generally considered as those that cannot be extracted with aqueous and organic solvents. However, a more precise definition has been provided by Skidmore et al:

> "A bound xenobiotic residue is a residue associated with one or more classes of endogenous macromolecules. It cannot be dissociated from the natural macromolecule using exhaustive extraction or digestion without significantly changing the nature of the associated endogenous macromolecules."

When studying bound-pesticide residues using radiolabeled pesticides, it is important to differentiate the bound residue containing the labeled xenobiotic or its metabolite from the "natural label." Natural labeling occurs when $^{14}CO_2$ is released from the mineralized pesticide and is incorporated into the plant cell wall. Natural labeling in plants has been observed with several pesticides.

Furthermore, it is important to know the precise position of the label on the pesticide molecule so that the site of pesticide incorporation into the cell wall can be determined. Digestive treatment with different enzymes such as cellulase, collagenase, pepsin, amylase, and proteases can aid in identifying the nature of pesticide incorporation. On the basis of reports in the literature, it appears that xenobiotics are incorporated randomly into different cell wall components; however, little is known about the type of linkages involved in this binding.

There is concern about the bioavailability of bound pesticides from plant residues. Phanerochaete chrysosporium mineralized bound chloroaniline and 2,4-dichlorophenol, indicating that these compounds may become bioavailable. The ability of animals to release xenobiotics bound to plant residues is unknown. Experiments using a "simulated stomach" demonstrated that pesticides were released from plant residues, but only when high concentrations of bound pesticide residues were used. In comparison, only low concentrations of bound pesticide residues are typically present in plant residues. However, the biological relevance of typically low concentrations of bound pesticide residues is not known. Presently, the U.S. Environmental Protection Agency requires no characterization of bound pesticide residues if concentrations are less than 0.05 ppm of the parent equivalents or 10% of the total pesticide residue. If concentrations exceed these levels, determination of the bioavailability based on "simulated stomach" experiments is required. The toxicological nature and bioavailability of bound xenobiotic residues requires continued research to fully assess its impact on human health and the environment.

Bioremediation and Pesticide Metabolism in the Rhizosphere

The ligninolytic fungus P. chrysosporium oxidizes the insecticide lindane by a putative cytochrome P450 enzyme. There is potential to exploit P. chrysosporium along with other indigenous microflora to mineralize lindane or convert it to volatile metabolites. Bioremediation of lindane-contaminated soil with P. chrysosporium is possible. The advantages of using filamentous fungi for bioremediation include the following: fungi are in direct contact with solid, liquid, and vapor portions of the soil, fungi are capable of transforming a large number of structurally dissimilar compounds, fungi are able to withstand toxic effects of many xenobiotics, and fungi release metabolites, making the metabolites available for further degradation by other microorganisms.

Use of Enzymes in Bioremediation

Bioremediation is the use of microorganisms, plants (often called phytoremediation), or biologically active agents to degrade, sequester, or conjugate environmental pollutants. Advantages of bioremediation include ease and timing of application, ability to target specific pollutants, decreased sludge volume, and decreased ecological hazard. There is potential to use enzymatic treatment in bioremediation, and this technology is currently at the laboratory stage of development. Advantages of enzymatic treatment over microbial bioremediation include no acclimation phase, use over a wider range of environmental conditions (pH, moisture, temperature), effectiveness at high and low pollutant concentrations, movement of enzymes readily into soil micropores and their protection from inactivation, and little effect of inhibitors of mi- crobial metabolism on enzymes. The disadvantages of enzymatic treatment in bioremediation include the high cost of isola- tion and storage, the difficulty in maintaining enzyme stability, the requirement for expensive cofactors, and the lack of xenobiotic mineralization.

The use of isolated enzymes to metabolize pesticides is not new. For example, enzymes from crude Pseudomonas cell extracts immobilized on glass beads, hydrolyzed 95% of parathion (10 to 250 ppm) from waste-water. The same enzyme preparation hydrolyzed parathion at 2,500 ppm in soil and was also effective in hydrolyzing other organophosphate insecticides. Microbial enzymes with potential for pesticide metabolism include oxidoreductases, hydroxylases, amidases, and esterases. However, enzymatic treatments are not ideal for complete xenobiotic mineralization because mineralization usually requires many enzymes and several cofactors such as NAD(P)H and FAD. Oxidoreductases, such as laccase, tyrosinase, and horseradish peroxidase, can be used to decontaminate soil and water. These enzymes oxidize the substrate to free radicals, which are susceptible to chemical coupling, forming oligomers. For example, oligomer formation reactions can take place between humic acid and xenobiotics, resulting in the polymerization of the substrate to soil, as was observed with 2,4-dichloro- phenol. In another experiment, horse- radish root tissue and hydrogen peroxide (an electron acceptor) decontaminated water containing 850 ppm of 2,4-dichlorophenol and other chlorinated phenols. Depending on the concentration of hydrogen peroxide, up to 100% of the contaminants were removed by polymerization. Furthermore, horseradish root tissue contributed to the irreversible binding of 2,4-dichlorophenol to soil.

For enzymatic treatment to be effective in bioremediation, the enzymes must be stabilized. The most effective way to stabilize enzymes is by immobilization. Immobilization can be accomplished by enzyme linkage to organic or inorganic solid supports by adsorption on solid surfaces such as glass, entrapment in polymeric gels, encapsulation, or intermolecular cross-linking. Although preparing supports can be time-consuming and expensive, the support can generally be reused. Enzymatic treatment holds great promise in bioremediation of contaminated soil and water.

Pesticide Degradation in the Rhizosphere

Chemicals released by plants may enhance xenobiotic degradation, and it may therefore be beneficial to use plants in the remediation of contaminated soils. There are three general mechanisms by which the rhizosphere may act to enhance cometabolism of anthropogenic contaminants. First, the rhizosphere may allow selective enrichment of degrader organisms that have densities too low to significantly degrade xenobiotics in root-free. Second, the rhizosphere may enhance growth-linked metabolism or stimulate microbial growth by providing a natural substrate when the concentration of xenobiotics is low or unavailable. Finally, the rhizosphere is rich in natural compounds that may induce cometabolism of xenobiotics in certain microorganisms that carry degradative genes or plasmids. This may permit initial degradation of xenobiotics that would otherwise be unavailable as carbon sources.

Rhizosphere effects on xenobiotic biotransformation have been studied for a variety of compounds, although the mechanisms by which certain plants enhance biodegradation are still poorly understood. Differences in plant tolerance to phytotoxic compounds in soils may be related to the plants' ability to induce microorganisms that will detoxify these xenobiotics in the soil environment. Research on phytoremediation, through trial and error, has focused on densely rooted, fast-growing grasses and plants, such as Brassica sp., with fine root systems. Mulberry and poplar trees have been used successfully in the phytoremediation of chlorophenols and chlorinated solvents such as trichloroethylene.

Salicylic acid, flavonoids, and monoterpenes are structurally analogous to many anthropogenic compounds in that they are small, mobile chemicals that are amenable to cellular uptake and may interact through signal transduction pathways to induce the production of specific degradative enzymes. For example, salicylic acid was used in tomato fields in irrigation water to promote rhizosphere bacterial growth. Salicylic acid is an effective inducer of many different enzymes that may be involved in the cometabolism of xenobiotics such as polyaromatic hydrocarbons (PAHs) and PCBs. Degradation of PAHs and PCBs probably evolved in a modular fashion by gene operon recruitment. The salicylic acid-inducible toluene monooxygenase gene, TOM, was isolated from Burkholderia cepacia and introduced into P. fluorescens. The rhizosphere of wheat was inoculated with this transformed P. fluorescens, resulting in enhanced degradation of TCE.

Pseudomonas putida G786 hydroxylates the terpenoid camphor by a monoxygenase P450CAM located on a plasmid. Other P450 enzymes have been implicated in terpene degradation; however, it appears that terpene-induced P450s do not have broad substrate specificity. Plants have been used in the phytoremediation of PAH-contaminated soils, suggesting the involvement of rhizosphere microorganisms in PAH degradation. A microbial community–based approach may be useful for screening different plant chemicals to find inducers of xenobiotic-degrading enzymes.

Reductive Dehalogenation in the Rhizosphere

Reductive dehalogenation is the only significant mechanism for the breakdown of halogenated aromatic, aliphatic, and heterocyclic compounds like PCBs, TCE, hexachloro-benzene, and halogenated pesticides such as heptachlor and aldrin. Reductive dehalogenation enzymes have broad substrate specificities. There are two principal mechanisms of RDE. The first process is cometabolic RDE, which yields no energy for the organism. The second mechanism is halorespiration, where organohalides act as terminal electron acceptors and adenosine triphosphate is generated. In cometabolic RDE, organohalides are not used as terminal electron acceptors. Generally, anaerobic respiration is relatively inefficient in that electrons produced during substrate oxidation lose energy in the electron transport chain. In cases of excess substrate, high-energy electrons, "hot electrons," accumulate, creating a redox potential imbalance. However, microbes may gain protection from "hot-energy" electrons by halo-scavenging, thereby providing an advantage to microbes that conduct halo- respiration.

In theory, certain microsites within the rhizosphere are favorable (anaerobic conditions, low redox potentials, and available electron acceptors) for RDE thus fascilitating the transformation of halogenated compounds. Although O_2 does not inhibit, RDE is generally an anaerobic process. Spatial and temporal heterogeneity in O_2 distribution in the rhizosphere environment usually (but not always) provides microbes with localized environments that are anaerobic and have low redox potential, thereby favoring RDE reactions. Moreover, most of the terminal electron acceptors, such as nitrate, ferric iron, sulfate, CO_2, and quinones, are abundant in the rhizosphere. The addition of quinones can enhance both the capacity and the rate of microbial reduction of contaminants. In RDE, quinones and semiquinones provide reducing power and protons, which transfer excess electrons to organohalides.

The bioavailability of hydrophobic contaminants determines the rate of xenobiotic transformation and mineralization. For example, the rhizosphere may increase the bioavailability of lipophilic polyhalogenated aryl halide contaminants otherwise unavailable for RDE. Concomitantly, the bioavailability of the hydrophilic intermediates of organohalide degradation will decrease, thus reducing further degradation. Further research is needed to fully characterize the role of the rhizosphere in halide degradation and the biotransformation of xenobiotics.

Effects of Herbicides on Crop Production

Food production started from time immemorial, with the geometric increase in population of human beings the need to increase food production to feed large number of people led to the advent of technologies like machines, herbicides among others. New production technologies designed to revamp the agricultural sector and boost agricultural production has led to marked increase in crop yields. The issue of providing adequate food supply to meet requisite demand in Nigeria has been topical for a number of years, problems of food production and distribution has been elaborately analyzed with a variety of policy recommendations, among which is the use of agrochemicals not only to increase food production but to reduce food waste and hopefully enhance food producers income. The conventional methods of raising farm productivity since the World War II has centered on employing the use of externally acquired inputs like fertilizers and protection chemicals among others.

Agrochemicals refer to substances used to help manage an agricultural ecosystem, or the community of organisms in a farming area. Agrochemicals are important agricultural inputs to protect crops from diseases, pests and weeds. The uses of agrochemicals contribute not only to healthy growth of crops and animals but also to improve farm work efficiency and stable supply of tasty agricultural produce. Agricultural chemicals include fertilizers, pesticides, herbicides, rodenticides to mention but just a few used to eliminate the presence of living things that causes injury or diseases to crops and to improve production. Although many kinds of chemicals are used in agriculture, they can be categorized into simple groups according to the functions they performed. This includes insecticides, herbicides, fungicides, molluscides, and rodenticides, just to mention but a few.

Herbicides use is not limited to crop production alone, its use in animal production include antibiotics administered either by injection or combined with feed; to control infectious diseases and parasites that often arise when animals are raised under extremely crowded conditions. In similar way to tractors, ploughs and other implements, herbicides have now become an integral part of the complex word of technical inputs required for modern agricultural production and are accepted as a standard tool of the trade by farmers throughout the world. For several years humans have utilized herbicides to protect their crops. The first known agrochemical was sulfur-dusting used in ancient summer about 4,500 years ago in ancient Mesopotamia. By the 15th century, toxic chemicals such as arsenic, mercury and lead were being applied to crops to kill insects. In the 17th century, nicotine sulfate was extracted from tobacco leaves for use as an insecticide. The 19th century saw the introduction of more natural agrochemicals; pesticides and pyrethrum, which are derived from chrysanthemums, and rotenone, which is derived from the roots of tropical vegetables.

In the 1940s manufacturers began to produce large amounts of synthetic herbicides and their use became widespread. Some sources consider the 1940s and 1950s to have been the start of the agrochemical era. According to Miller, agrochemical use has increased 50 fold, since 1950 and 2.3 million tonnes (2.5 million short tonnes) of industrial pesticides are now used each year. Seventy-five percent (75%) of all herbicides in the world are used in developed countries however; its use in developing countries is increasing. Many herbicides can be grouped into chemical families. Prominent pesticide families include organocholrines, organophosphates and carbamates. Organochlorine hydrocarbons (DDT) could be separated into dichlorodiphenylethanes, cyclodiene compounds, and other related compounds. Prominent families of herbicides include pheoxy and benzoic acid herbicides (2, 4-D) tyriazines (atrazine), Ureas (diuron), and Chloroacetanilides (alachlor). Many commonly used herbicides are not included in these families, including glyphosate.

The primary objective of agriculture is to produce a reliable supply of food for the increasing world population. It plays an important role as a source of food, of foreign exchange for imports and of exports, employment and imports for other sectors in the economies of many industrialized as well as developing countries. Crop production has however, been threatened by various insects, weeds among others. The FAO has estimated pre-harvest crops losses due to weed infestation, plant diseases; arthropods (largely insects and termites) to be around 30 to 35%, and post-harvest losses (grain storage, etc.) amounted to an additional 10 – 20%.

In order to arrest these problems, chemical weed control has become an increasingly necessary operation in the consistent and economic production of crops. In addition to agricultural diversification and yield optimization, chemical weed control has formed an integral part of the policies of many governments world over. With benefits of herbicides control ranking high, negative effects on the environment and human health generated mainly by lack of knowledge regarding safety parameters on the part of the user has made herbicides use in agriculture one of today's most controversial issues.

Agricultural chemicals have significantly increase crop yield in the short term by limiting damage by pest, competition for water and nutrients from weeds and by providing large amounts of nutrients in a form that is easily available to plants. In the long run these processes can lead to serious depletion of soils because the natural processes of converting organic matter and the balance of microorganisms in the soils have been disrupted. Herbicides used against control of grasses can kill beneficial insects like lady bugs, aphids, among others. When the beneficial insects are gone, there is no natural control over the pest, so their populations can increase much more quickly after the initial application, requiring further applications of pesticides to control the original pest. Herbicides can kill butter flies, moths, spiders, and bees which play other roles in the environment such as pollinating plants.

With the increasing awareness concerning health hazards associated with herbicides on the part of consumers, environment and farming groups, there has been a consequent demand for more stringent regulatory measures in the development of environmentally safe agrochemical formulations. However, mankind will still continue to use chemicals to produce adequate supplies of food and feeds, while trying to minimize associated risks to human health and the environment. Risk is an In most countries, herbicides must be approved for sale and use by a government agency. For example, in the United States, the Environmental Protection Agency (EPA) does so. Studies must be conducted to indicate whether the material is safe to use and effective against the intended herb. During the registration process, a label is created. The label contains directions for use of the material. Based on acute toxicity, herbicides are assigned to toxicity class. Some agrochemicals are considered too hazardous for sale to the general public and are designated restricted use agrochemical. Only certified applicators, which have passed an examination may purchase or supervise the application of restricted use chemicals.

Alternatives to the use of pesticides are available and it include methods of cultivation, use of biological pest controls (such as pheromones and microbial pesticides), genetic engineering, and methods of interfering with insect breeding.

Methods of alternative weed control are becoming increasingly popular and are often safer than traditional herbicides. Cultivation practices include polyculture (growing multiple types of plants), crop rotation, planting crops in areas where the pests that damage them do not live, timing planting according to when pests will be least problematic and use of crops that attract pests away from the real crop. In some developed countries, (USA in particular) farmers have had success controlling insects by spraying with hot water. Release of other organisms that fight the pest is another example of an alternative to pesticide use. These organisms can include natural predators or parasites of the pests. With the increasing awareness concerning health hazards associated with herbicides on the part of consumers, environment and farming groups, there has been a consequent demand for more stringent regulatory measures in the development of environmentally safe agrochemical formulations. However, mankind will still continue to use chemicals to produce adequate supplies of food and feeds, while trying to minimize associated risks to human health and the environment. Risk is an intrinsic part of life and must be weighed against the benefits likely to result from any particular action.

There is a lack of effective disposal methods; this depends on the package design and toxicity of herbicides. Widespread use of empty containers, lack of quantification and documentation of wastes, poor storage conditions for obsolete stocks, especially in developing countries. The package design should also discourage the widespread reuse of herbicide containers, which is a common practice in developing countries, and should encourage reductions in waste herbicides. It is therefore, desirable to ensure safe and effective use of herbicides by increasing awareness, training and the dissemination of relevant information, and by enacting legislation of control sales, distribution, use, production, formulation and disposal. There is further need for research into alternative herbicides and the deployment of herb – management strategies. The past decades have witnessed a dramatic change in agriculture with food production soaring due to the Green Revolution.

The Green Revolution entailed the use of improved technologies; particularly the breeding of high yielding food crop varieties, the expansion of irrigation, mechanization, specialization and the use of chemical fertilizers and pesticides. However, the Green Revolution has been criticized for its adverse human health and environmental impacts. For example, agricultural intensification through excessive and inappropriate use of chemical fertilizers and pesticides, has polluted water bodies and degraded soils, led to biodiversity loss by killing beneficial plants, insects and other wildlife and in some cases poisoned farm workers, to mention but just a few. Sustainable agriculture has emerged as an alternative agricultural system that addresses the many constraints faced by poor-resource farmers and at the same time ensures environmental sustainability.

For agricultural production to be sustainable crop and livestock production must increase without an increase in the negative environmental impacts associated with agriculture, which means large increases in the efficiency of nitrogen, phosphorus and water use and integrated pest management that minimizes the need for toxic pesticides. Nutrient-use efficiency is increased by better matching temporal and spatial nutrient supply with plant demand. Applying fertilizers during periods of greatest crop demand, at or near plant roots, and in small and frequent applications all have the potential to reduce losses while maintaining or improving yield and quality. Such precision of agriculture has typically been used in large-scale intensive farming, but is possible at any scale and under any conditions giving the use appropriate diagnostic tools. Strategies that synchronize nutrient release from organic sources with plant demand are also needed.

Multiple cropping systems using crop rotations or intercropping may improve pest control and increases nutrient and water use efficiency. Agroforestry, in which trees are included in a cropping system, may reduce erosion, provide firewood and store carbon. Landscape-scale management holds significant potential for reducing off-site consequences of agriculture. Individual farms, watersheds and regional planning can take advantage of services provided by adjacent natural, semi-natural or restored ecosystems. Trees and shrubs planted in buffer strips surrounding cultivated fields decrease soil erosion and take up nutrients that otherwise would enter surface or ground waters. Buffer zones along streams, rivers and lakeshores can decrease nutrient and loading from cultivated fields or pastures. Crop pollination can be provided by insects and other animals living in nearby or buffer strips, whereas other organisms from these habitats, such as parasitoids, can provide effective control of many agricultural pests. Buffer strips can also manage to reduce inputs of weeds and other agricultural pests.

Environmental sustainability is typically achieved through protecting, recycling, replacing and maintaining the natural resources base such as land, water and wildlife that contribute towards conservation of natural capital. While synthetic fertilizers can be used to supplement natural inputs, they are applied on needs basis. Under sustainable agriculture, synthetic chemicals known to harm soil organisms, soil structure and biodiversity should be avoided or reduced to minimum used.

Kassie and Zikhali social sustainability relates to the quality of life those who work and live on the farm, as well as those in the surrounding communities. It includes ensuring equitable revenue or returns to different stakeholders of the agricultural production chain. In context of unemployment sustainable agriculture can promote sharing of agricultural value added by more members of the community through more extensive use of available labour, at least for some techniques, thus contributing to social justice and cultural cohesion. Fair treatment of workers and choosing to purchase supplies locally rather than more distant markets. Economic sustainability should also be ensured, farms that are not economically viable are replaced by alternative uses of land that are more profitable. This can improve the economic viability of a farm in a number of ways. In the short term, improving soil management and crop rotation can increase yields, while in the medium and long term, improved soil quality and water availability as well as other environmental benefits from sustainable practices, may raise the value of the farm and provide for payment for environmental services. Economic viability can also be achieved through, for example, reducing machinery, chemical fertilizer and pesticide costs (for farmers who afford these inputs), depending on the specific characteristics of the production system. Of course, economic sustainability is also conditioned by many factors aside from crop production methods, for example household characteristics such as managerial ability, institutions, and infrastructure and market access among others.

Nigeria government should adopt policies or strategies as it has been done in other countries including Australia, Canada, European Union countries, Japan, Norway, Switzerland and United States which instituted various forms of green payments, that is payment to farmers who adopt sustainable farming practices. Other policy options include taxes, removal of subsidies, a tax on fertilizers or pesticides, or removal of subsidies for these inputs would discourage excessive use.

The pursuit of sustainable agriculture will also require substantial increase in knowledge-intensive technologies that enhance scientifically sound decision making at the field level. The earlier paradigm of science being developed at the international or perhaps national level and then disseminated to farmers should be replaced by an active exchange of information among scientists and farmers. Scientists in Nigeria who understand the ecosystems, human culture and demand on local agricultural systems must be actively trained, promoted and brought into the international scientific community.

Methodology

Simple random sampling technique was adopted in collecting data. Data were collected through primary and secondary sources. Primary source was by survey using structured questionnaire administered on 80 respondents (farmers that used herbicides) selected randomly from four council wards. Secondary data were collected from books, journals, proceedings and documents. Data collected were analyzed through descriptive statistics.

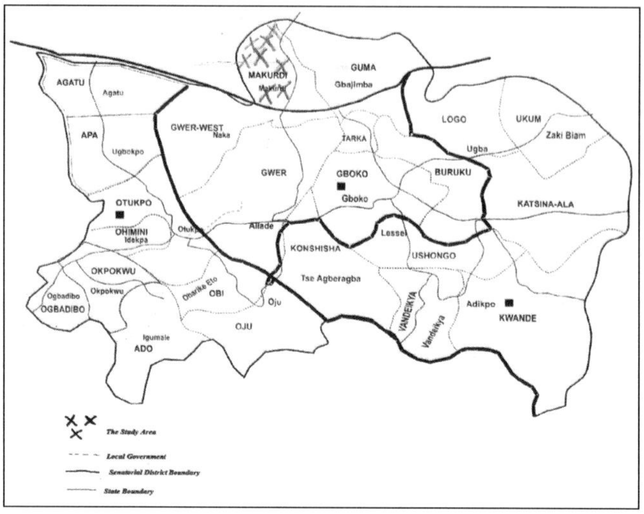

Table: Distribution of respondents according to time in a day when herbicides are applied.

Time	Frequency	Percentage
Morning	60	75.0
Evening	14	17.5
Afternoon	6	7.5
Total	80	100

The results in showed that 75.0% of the respondents applied herbicides in the morning. This could be attributed to the fact that most of the people used for application of herbicides in the villages do not wear any head mask before applying herbicides, because they do not wear head mask or boot, the best time to apply herbicides is in the morning. Apart from that, herbicides can easily evaporate leaving nothing to the farmer especially in tropics where the temperature in our environment is usually very high; its application in the afternoon allows a reasonable quantity of it to evaporate, therefore if applied in the afternoon the aim of the application is not achieved. To ensure proper utilization of herbicide it is better applied in the morning or evening when the temperature is low. The application of the herbicides in the morning make it to be absorbed easily by the herbs better than when it is applied at any other time of the day.

Table: Distribution of respondents according to types of herbicides used.

Types of Herbicides	Frequency	Percentage
Non-selective	88	110.00
Systemic	85	106.25
Selective	43	53.75
Contact	15	18.75
Insecticides	7	8.75

Results in table showed that majority 110.0% used non-selective herbicides. Non-selective herbicides are used in non-crop situation; land clearing and industrial weed control, railway tracks, and forestry sites. Benue State belongs to vegetation belt that is characterized by high grasses (grassland area) therefore; farmers in their attempts to clear grasses/herbs for farming make use of non-selective herbicides. The use of herbicides for clearing herbs is faster, and energy conserving than the use of any other physical means especially hands. The use of herbicides for clearing herbs in preparation for farming is very common in Benue State especially in the last 8 years. Most farmers engage hired people who use herbicides and clear herbs which are very economical. If physical weeding is employed it may take farmers many months to clear preparatory to farming especially for those with large farm, however with the use of herbicides a few days are used to cover the same piece of land that could have taken the farmers many months to clear by weeding. The composition of herbicides when it is to be applied among crops and grasses/herbs is also a major problem because most of the people involved in spraying do not undergo any form of training to make them understand herbicides composition and application.

Table: Distribution of respondents according to health hazards associated with the use of herbicides.

Hazards	Frequency	Percentage
Fatigue	80	100
Eye problem	76	95.0
Skin problem	70	87.5
Dizziness	32	40.0
Vomiting	9	11.25

Table showed that 100% of the respondents stated that they had fatigue while applying herbicides. This could be attributed to attempts to cover large hectares of farmland within a short period of time and the use of faulty spraying equipment which cause linkages. Majority of the people hire spraying equipment for application of herbicides. Hiring of sprayers is time bound therefore, the farmers are eager to complete their job and returned the sprayers to the owners for fear of been sub charged. If it is a hired person, their attempts to cover large area of farmland often result to spraying fast which mostly results to fatigue, this also leads to haphazard application of herbicides which very often result to the poor death of the herbs. Furthermore, the poor death of the herbs after spraying herbicides on farmland could require a farmer to use other means in further clearing the herbs that were left over as a result of poor application of herbicides because most of the farm activities are done manually. The effective use of herbicides depends to a great extent on how they are applied. More so, much of the equipment used in developing countries is poor and some designs have proved unsuitable for use.

Table: Distribution of respondents based on precautions observed when applying herbicides.

Precautions	Frequency	Percentage
Read label before use	80	100
Wear head mask	64	80.00
Wear overall or apron	39	45.00
Wear boot only	38	47.50

Results in table above showed that 100% of the respondents read the instruction rebel of the herbicides before applying it. This is an indication that most people that use herbicides can read and write. Majority of the farmers employ the services of commercial herbicide sprayers who sometimes go out from one house to another looking for farmers to hire them to do the job. Apart from farmers, schools (primary and secondary) and religious places make use of herbicides in clearing herbs in and around their premises. One thing with the commercial herbicide sprayers is that most of them are very familiar with the herbicides commonly used in the area and they handle these herbicides as if they were not poisonous, a situation that will present a very dangerous future for them but most of them are not aware of the long time effect herbicides will have on their health.

Table: Distribution of respondents according to direction use when applying herbicides.

Direction	Frequency	Percentage
Following the direction of wind	66	82.5
Wind direction not considered	13	16.25
Against the direction of wind	1	1.25
Total	80	100

Results in indicated that 82.5% applied herbicides following the direction of wind. Direction of wind when applying herbicides is very important, at any time herbicides or agrochemicals are to be applied the direction of wind has to be considered. Herbicides are usually applied against the direction of wind. However most of the people in the rural area that apply herbicides are not trained. Majority of them saw others people making use of sprayers; they bought theirs and started using it either for personal or commercial purpose without knowing other requirements for application of herbicides. That is why most of the people involved in the application herbicide seem not very effective in carrying out the job. In addition to poor spraying, some of them cannot identify whether a sprayer is very good or bad at time of buying it and these constitute major problems in effective application of herbicides.

Table: Distribution of respondents according to factors militating against the use of herbicides.

Factors	Frequency	Percentage
High cost	80	100
Contaminate water sources	10	12.5
Killing of non-targeted Organisms	2	2.5
Non-availability	2	2.5

Results in the table depicted 80% high costs of herbicides. The high cost of herbicides is as a result of middle men's involvement in the sale and distribution of herbicides. The middle men in their attempt to maximize profit buy in large quantities and hike the price of herbicides almost beyond the average price making it difficult for the peasant farmers to buy. Apart from that most of the peasant farmers' farmland is not big enough to engage people outside their family as that will amount to waste of their little resources. So to those groups of farmers whose farmland is small it is not economical to use herbicides in any way on their farm as it will only amount to economic waste.

In conclusion, the advent of herbicides has led to increase in crop yield and reduce drudgery among farmers; little time is now taken to clear herbs especially before farming. Most farmers applied herbicides in the morning the commonly used herbicide is non- selective. Furthermore, the common health hazard associated with application of herbicides is fatigue, though people employed for application of herbicides read instruction label before its application, they do not observe other instructions like wearing of mask and the right direction during herbicides application. High cost is one of the problems militating against the use of herbicides. Herbicides are good especially for clearing of weed around our environment; they cause a lot of damage by killing the non-targeted beneficial insects thereby creating more problems in the near future. It is recommended that Agricultural Development Project (ADP) should organize training for farmers on application of herbicides and farmers should form associations so as to pull resources together, buy herbicides directly from the distributors in large quantity and disburse the herbicides among themselves to prevent been exploited especially the hike in price by the middle men.

Environmental Effects of Herbicides

Some substantial benefits can be gained through the use of herbicides to manage unwanted vegetation. Compared with alternative means of weed control, such as mechanically weeding by hand or machine, herbicides are less expensive, often safer (especially in forestry), faster, and sometimes more selective.

However, if herbicides are not used properly, damage may be caused to crop plants, especially if too large a dose is used, or if spraying occurs during a time when the crop species is sensitive to the herbicide. Unintended but economically important damage to crop plants is sometimes a consequence of the inappropriate use of herbicides.

In addition, some important environmental effects are associated with the use of herbicides. These include unintended damage occurring both on the sprayed site, and offsite. For example, by changing the vegetation of treated sites, herbicide use also changes the habitat of animals such as mammals and birds. This is especially true of herbicides use in forestry, because biodiverse, semi-natural habitats are involved. This is an indirect effect of herbicide use, because it does not involve toxicity caused to the animal by the herbicide. Nevertheless, the effects can be severe for some species. In addition, not all of the herbicide sprayed by a tractor or aircraft deposits onto the intended spray area. Often there is drift of herbicide beyond the intended spray site, and unintended, offsite damages may be caused to vegetation. There are also concerns about the toxicity of some herbicides, which may affect people using these chemicals during the course of their occupation (i.e., when spraying pesticides), people indirectly exposed through drift or residues on food, and wildlife. For these and other reasons, there are many negative opinions about the broadcast spraying of herbicides and other pesticides, and this practice is highly controversial.

The intention of any herbicide treatment is to reduce the abundance of weeds to below some economically acceptable threshold, judged on the basis of the amount of damage that can be tolerated to crops. Sometimes, this objective can be attained without causing significant damage to non-target plants. For example, some herbicides can be applied using spot applicators or injectors, which minimize the exposure to non-pest plants and animals. Usually, however, the typical method of herbicide application is some sort of broadcast application, in which a large area is treated all at once, generally by an aircraft or a tractor-drawn apparatus.

An important problem with broadcast applications is that they are non-selective—they affect many plants and animals that are not weeds—the intended target of the treatment. This is especially true of herbicides, because they are toxic to a wide variety of plant species, and not just the weeds. Therefore, the broadcast spraying of herbicides results in broad exposures of non-pest species, which can cause an unintended but substantial mortality of non-target plants. For example, only a few species of plants in any agricultural field or forestry plantation are abundant enough to significantly interfere with the productivity of crop plants. Only these competitive plants are weeds, and these are the only target of a herbicide application. However, there are many other, non-pest species of plants in the field or plantation that do not interfere with the growth of the crop plants, and these are also affected by the herbicide, but not to any benefit in terms of vegetation management. In fact, especially in forestry, the non-target plants may be beneficial, by providing food and habitat for animals, and helping to prevent erosion and leaching of nutrients.

This common non-target effect of broadcast sprays of herbicides and other pesticides is an unfortunate consequence of the use of this non-selective technology to deal with pest problems. So far, effective alternatives to the broadcast use of herbicides have not been discovered for the great majority of weed management problems. However, there are a few examples that demonstrate how research could discover pest-specific methods of controlling weeds that cause little non-target damage. These mostly involve weeds introduced from foreign countries, and that became economically important pests in their new habitats. Several weed species have been successfully controlled biologically, by introducing native herbivores of invasive weeds. For example, the klamath weed (Hypericum perforatum) is a European plant that became a serious pasture weed in North America, but it was specifically controlled by the introduction of two species of herbivorous leaf beetles from its native range. In another case, the prickly pear cactus (Opuntia spp.) became an important weed in Australia after it was introduced there from North America, but it has been successfully controlled by the introduction of a moth whose larvae feed on the cactus. Unfortunately, few weed problems can now be dealt with in these specific ways, and until better methods of control are discovered, herbicides will continue to be used in agriculture, forestry, and for other reasons.

Most herbicides are specifically plant poisons, and are not very toxic to animals. (There are exceptions, however, as is the case with the herbicide paraquat.) However, by inducing large changes in vegetation, herbicides can indirectly affect populations of birds, mammals, insects, and other animals through changes in the nature of their habitat.

For example, studies in Britain suggest that since the 1950s, there have been large changes in the populations of some birds that breed on agricultural land. These changes may be partly caused by the extensive use of herbicides, a practice that has changed the species and abundance of non-crop plants in agroecosystems. This affects the structure of habitats, the availability of nest sites, the food available to granivorous birds, which mostly eat weed seeds, and the food available for birds that eat arthropods, which rely mainly on non-crop plants for nourishment and habitat. During the time that herbicide use was increasing in Britain, there were also other changes in agricultural practices. These include the elimination of hedgerows from many landscapes, changes in cultivation methodologies, new crop species, increases in the use of insecticides and fungicides, and improved methods of seed cleaning, resulting in fewer weed seeds being sown with crop seed. Still, a common opinion of ecologists studying the large declines of birds, such as the gray partridge (Perdix perdix), is that herbicide use has played a central but indirect role by causing habitat changes, especially by decreasing the abundance of weed seeds and arthropods available as food for the birds.

Similarly, the herbicides most commonly used in forestry are not particularly toxic to animals. Their use does however, cause large changes in the habitat available on clear-cuts and plantations, and these might be expected to diminish the suitability of sprayed sites for the many species of song birds, mammals, and other animals that utilize those habitats.

Modern, intensively managed agricultural and forestry systems have an intrinsic reliance on the use of herbicides and other pesticides. Unfortunately, the use of herbicides and other pesticides carries risks to humans through exposure to these potentially toxic chemicals, and to ecosystems through direct toxicity caused to non-target species, and through changes in habitat. Nevertheless, until newer and more pest-specific solutions to weed-management problems are developed, there will be a continued reliance on herbicides in agriculture, forestry, and for other purposes, such as lawn care.

After a herbicide is applied, whether pre-plant incorporated, pre-emergence, or post-emergence, inside a polyhouse/structure, or to any other site, it has been introduced into the environment. Applicators need to ask themselves a few important questions:

1. Will the herbicide remain where it was applied or will the pesticide become mobile in the environment?
2. How long will the herbicide remain viable or effective?
3. What effect could the herbicide have on non-target plants, animals, or other things in the environment?

To answer these questions, you must understand how herbicides move in the environment and the chemical properties that control movement. There are four basic chemical characteristics that control herbicide/pesticide movement in the environment: solubility, adsorption, persistence and volatility.

Solubility is a measure of the ability of a pesticide to dissolve in a solvent, usually water. The greater the solubility, the more readily the pesticide dissolves. Pesticides that are easily dissolved in water can move with water. Highly soluble pesticides are more likely to move through the soil and into groundwater or into surface waters, causing harm to unintended sites, plants and animals, including humans.

Adsorption is the ability of a pesticide to bind with soil particles. Adsorption occurs because the pesticide has an electrical attraction to the surface electrical charge of a soil particle, generally organic matter or clay particles. A pesticide that adsorbs to soil particles is less likely to move from the application site.

Persistence is the ability of a pesticide to remain in its original form, active and viable, before breaking down chemically to become inactive. A common measure of persistence in chemicals is referred to as the half-life. Half-life is the time it takes for half the original amount of chemical applied to break down. The longer the reported half-life of a chemical or pesticide, the more persistent the chemical or pesticide is in the environment. Sometimes, persistent pesticides are desirable because they will provide long-term pest control and reduce the need for repeated applications. However, persistent pesticides can also cause later problems to unintended sites, plants, animals or humans if the persistent pesticides are also mobile in the environment. If you are using a persistent pesticide, it is very important to prevent unintended consequences due to improper handling, drift, runoff, erosion or leaching.

Volatility is a measure of the tendency of a pesticide to turn into a gas or vapor. Some pesticides are more volatile than others. Pesticides tend to volatilize more readily when temperatures are high, winds are high, and relative humidity is low at the application site. Pesticide movement as a gas or vapor is also known as "drift".

Pesticide degradation occurs in three basic ways:

- Microbial action: Chemical breakdown or degradation of pesticides by soil microorganisms, such as fungi, bacteria, etc.
- Chemical degradation: Breakdown of pesticide chemical components by inorganic methods (not by living organisms).
- Photodegradation: Breakdown of pesticide chemical components by reaction with sunlight. This is why many pesticide application instructions require incorporation of the pesticide in the soil, away from direct sunlight.

How do Pesticides Move in the Environment?

Pesticides can move in the air, in water, and through the soil resulting in environmental damage and exposure to nontarget plants and animals. Applicators are responsible for damages resulting from off-target pesticide movement.

Pesticide Drift

Pesticide drift is the movement of pesticides through the air away from the intended target site. When pesticide drift occurs, it can damage crops and expose humans, domestic animals and wildlife. Drift can contaminate soil and water.

Pesticide movement in water usually is the result of either runoff from the application site to an unintended site or water body or leaching from the soil by water, moving outward and/or downward in the soil. This can cause unintended harm to plants or animals or contaminate surface water or groundwater.

Movement on or in objects includes such things as:

- Pesticide residues on equipment or clothing used by pesticide applicators. These residues can affect unintended plants, wildlife, livestock, pets and people.
- Pesticides that have adsorbed on soil particles that are subsequently moved to an unintended site by wind or water erosion.
- Pesticide residues on plants that are removed from site. This may be as plant parts, feed, seed or other plant-based products.

- Pesticide residues on or in animals that are treated by pesticides and moved to a new site. The residues can be in the meat, milk or fiber used by man, on their fur or skin, in their feces or other waste products, etc. Minimizing pesticide movement and subsequent unintended application and damage is part of the pesticide applicator's job.

Types of Pesticide Drift

There are two types of drift: vapor drift (chemical volatility) and particle drift. Vapor drift is the movement of pesticide vapors from the target area, carried by air. Particle drift is the movement of small spray droplets or dust from the target area, carried by air.

Factors Affecting Drift

Many factors influence the amount of particle spray drift. Of primary concern are spray droplet size and wind velocity, as they are the cause of most of the problems associated with spray drift. Temperature influences the volatility of pesticides. The size of the spray droplets determines how fast the droplets fall and how far the pesticide might drift. Small, lightweight droplets fall more slowly and have more time to drift. As droplet size increases, the potential for drift decreases. The greater the wind speed during a pesticide application, the greater the risk of pesticide drift. Consider wind direction when planning a pesticide application. Do not apply pesticides when the wind is blowing towards a susceptible crop, water body or other sensitive site. Winds are generally calmer in early morning or early evening. These are better times of day to apply pesticides. Low relative humidity and/or high temperature increase the evaporation rate of spray droplets. Spray drift is usually greater from aerial applications than from ground applications. Proper nozzle selection helps maintain uniform application by controlling both the amount of pesticide applied and the size of the pesticide droplets. Spray pressure influences the size of the droplets formed. Increased pressure produces smaller droplets, which are more susceptible to drift.

Pesticide Contamination and Water Resources

Pesticides can contaminate surface water and groundwater from both point sources and nonpoint sources. Chemicals on the ground surface can become groundwater contaminants if they are carried downward by recharge water. Leaching is the term for transport of pesticides downward or sideways through soil. The risk of Groundwater contamination is greater when pesticides are applied to gravelly or sandy soils. The closer the water table is to the land surface, the greater the possibility of contamination. Runoff and erosion moves pesticides into surface water bodies, such as streams or lakes.

Protecting Water Resources

Preventing Groundwater contamination is the pesticide applicator's responsibility. Always read, understand and follow directions and precautions on the product label. Use pesticides only when and where necessary and only in amounts adequate to control pests. There are both civil and criminal penalties for using pesticides in a manner inconsistent with label directions. Read, understand and follow the information and instructions on the pesticide label regarding disposing of pesticides and storing pesticides safely. Maintain records of pesticide applications, as required. Additional Groundwater protection methods, such as timing of irrigation, avoiding irrigation runoff and regularly inspecting and maintaining water wells, can help prevent groundwater contamination.

Chemigation

Chemigation is the application of agricultural chemicals, both pesticides and fertilizers, through a sprinkler system. Chemigation has the advantage that the correct amount of chemical can be applied to the crop at the appropriate time, the application is inexpensive, convenient, and field access is unnecessary.

Pesticide Effects on Non-target Organisms

The effects of pesticides on non-target organisms may involve direct and immediate injury or may be due to the long-term consequences of environmental pollution. Valuable non-target plants, bees and other beneficial insects, pets, livestock, and wildlife may be affected.

Phytotoxicity refers to plant injury caused by exposure to a chemical. Phytotoxic injury can occur on any part of a plant's roots, stems, leaves, flowers or fruits. Most phytotoxic injuries are due to herbicides that are persistent at the site of application. Persistent products may also injure succeeding crops. Damage to crops or other plants in adjacent areas is most often due to drift, although damage may sometimes be a consequence of surface runoff, particularly from sloping areas. Apply pesticides in the evening or early morning, when bees are not actively foraging. Beneficial insects, other than bees, can also be harmed by pesticides. Survey the insect population and use caution when applying pesticides. Keep pets out of treated areas during applications and cover pet food and water bowls. Applying a pesticide to a forage crop that is not listed on the label and then feeding the forage to livestock may result in illness or death of the animals. Pesticides can affect wildlife in many ways. They may kill wildlife, weaken wildlife, kill their food source or interfere with reproduction. Lethal effects are those that cause death directly by exposure to pesticides. Sublethal effects are those that do not kill outright, but those that interfere with survival and reproduction. Bioaccumulation or bioconcentration is the accumulation of persistent pesticides in the bodies of animals. Biomagnification is the accumulation of persistent pesticides in increasing concentration in animals as it moves up the food chain. Any application method or farming practice that allows considerable drift or runoff is potentially harmful to wildlife. Remember, if you use pesticides, you are responsible for knowing if an endangered or threatened species or their habitat may be affected by pesticide use in your area.

Effects of Herbicides on Nitrogen Cycling

In an attempt to minimize the production costs and environmental impacts that occur because of the use of N fertilizers, the adoption of management strategies that minimize the N losses in agroecosystems has been proposed. Such strategies include the efficient application of N at the moment the crop needs it most. N presents a great difficulty, however, because it is a highly reactive element that is subject to various changes in the soil and in plants. Most of the changes of N in soil are intermediated by the microbial community, which fluctuates in both time and space.

Therefore, besides properties such as pH, organic matter and CTC content, N dynamic is influenced by factors that affect the microbiota. These factors include aeration, water availability and the presence of toxic or stimulants substances to microorganisms.

The effects of herbicides on soil microbiota depend on the mechanism and other features of the product; type of metabolism of the microorganism (e.g., photolithotrophic, photoorganotrophic, chemolithotrophic, chemoorganotrophic), which determines if the metabolic routes that are affected by the herbicide are present in the microorganism; and also edaphic and climatic factors. Some herbicides, such as the paraquat, are toxic to biota in general. Others, such as glyphosate and glufosinate ammonium show low toxicity to nontarget organisms. In fact, an increase in microbial biomass after the application of these products has been observed in some situations. This result has been attributed to the products serving as a source of nutrients. It is worth stressing, however, that some studies have shown that glyphosate is toxic to some organisms, such as certain strains of nitrogenfixing bacteria (NFB). Besides affecting processes of the N cycle that are intermediated by the microbiota, the herbicides can change the metabolism of N inside the plant, depending on the mechanism of the product. The herbicide glufosinate ammonium, for example, inhibits the incorporation of NH_4^+ in organic compounds. This effect can cause a build-up of NH_4^+ in vegetal tissue. Because its ions are toxic to plants, part of the accumulated NH_4^+ can be exuded through the roots and/or converted into NH_3 and lost in the atmosphere, reducing the N availability in the system. The exit of N from the plant can increase the C/N relation of the formed stubble, thus reducing its mineralization rate. Some of the processes of the N cycle and the effect desiccant herbicides have over them are presented below.

Nitrogen Losses

Among the main forms of nitrogen loss in agroecosystems are removal by crop, the volatilization of soil ammonium, the gas loss of nitrogen oxides (NO_2, N_2O, NO) and elemental nitrogen (N_2), the lixiviation of soil nitrate leaving the root system's target range and NH_3 loss by the aerial parts of plants. Sutton et al. reported that losses of nitrogen by the aerial parts of vegetables can increase ammonium levels in the atmosphere, noting emission values ranging from 5-15 kg ha-1 year-1 for N-NH_3. This nitrogen output may be responsible for 15-20% of the total NH_3 gas emissions. The magnitude of nitrogen losses through the aerial parts of crops varies according to each species. Wetselaar & Farquhar estimated from research available in the literature that the total N losses through the aerial parts of wheat, rice, grassy sorghum, cotton and rye crops were within a range of 16 to 40, 48, 48, 21 and 42 kg ha^{-1} N, respectively. For barley crops, Schjoerring et al. found losses of 1-2 kg ha^{-1} $year^{-1}$ of N-NH_3 under three fertilization regimes with N. For corn, Francis et al. found losses of 45 to 81 kg ha^{-1} N, with application rates ranging from 50 to 300 kg ha-1 N. NH_3 losses through the aerial parts of sugar cane were indirectly estimated by Trivelin at around 50 kg ha^{-1} $year^{-1}$. This number represents half of the dose of N fertilizer applied to the crop sprouts.

Aside from its effects on various species, N losses can be linked to the developmental stages of plants, N availability in the soil, stomatal conductance, leaf temperature, NH_3 concentrations in the atmosphere, ammonium concentrations in the plant, the activity of the glutamine sinthetase enzyme and environmental stresses. Another important factor that affects not only the intensity but also the way in which ammonium gas is exchanged among leaves and the atmosphere is the ammonium compensation point, which was defined by Farquhar et al. When NH_3 concentrations in the atmosphere are below the ammonium compensation point of the plant, emission will occur through the leaves; for concentrations above, it is absorbed. Senescence is the developmental stage that shows the highest NH_3 emission potential. During this stage, there is an increase in protein degradation, and the liberated N is transferred to the glutamate. Afterward, the glutamate N is converted to NH_4^+ by the glutamate dehydrogenase (GDH) enzyme. GDH is an enzyme of N metabolism that usually reaches a higher degree of activity during senescence. NH_4^+ is available for transformation in the composts of N transport, such as glutamine and asparagine, and some of the ions are converted into NH_3 according to the following balanced reaction:

$$NH_4^+ \leftrightarrow NH_3 + H^+ \quad pKa = 9.2$$

When the pH is 9.2, 50% of the molecules are dissociated (NH_4^+), and 50% are in the molecular form (NH_3). In the cytoplasm, the pH is usually around 7.2. At this pH value, only 1% of the molecules are in the NH_3 form. However, it is important to stress that the NH_4^+ build-up in the cytoplasm can promote H^+ exit to the apoplast, transiently increasing the interior pH of the cell and, consequently, the NH_3 concentration. There is evidence that NH_3 transport is easier through membranes, and that it can be stomatally lost.

Effect of Herbicides

Herbicides are used in agricultural areas to induce the senescence of plants. The duration of the senescence phase deriving from herbicide application is very short, and the intervention may occur at any developmental stage. Therefore, it is possible that the behavioral standards of N are different from those observed in natural senescence conditions. Besides inducing senescence, glyphosate and glufosinate ammonium herbicides also affect the plant's N metabolism. The herbicidal action of glyphosate is attributed to its inhibition of the EPSPs (enolpyruvylshikimate phosphate synthase) enzyme that is responsible for one of the synthesis phases of the tryptophan amino acids phenylalanine and tyrosine. However, as a secondary effect, there is an inhibition in the synthesis of phenolic compounds deriving from those amino acids, which increases the activity of the PAL (phenylalamine ammonia lyase) enzyme due to a response effect. The PAL acts in the lyase of phenylalanine and tyrosine amino acids, resulting in the formation of phenolic acids and NH_4^+. The glufosinate ammonium herbicide inhibits the action of the glutamine sinthetase, which is responsible for the glutamate synthesis from ammonium and glutamine. This metabolic route is considered the main form of ammonium assimilation in organic composts. Consequently, the application of glufosinate ammonium results in endogenic NH_4^+ in the plant.

Therefore, the increase of NH_4^+ in plants desiccated with glyphosate and glufosinate ammonium should not be connected to the induction senescence alone. Damin et al. quantified N losses by Brachiaria decumbens after desiccation with glyphosate, glufosinate ammonium and paraquat, all of which promoted senescence. However, only the first two increased the losses of N in the system. Manderscheid et al. assessed the ammonium concentration in five weed species after applying glufosinate ammonium. They observed an NH_4^+ increase in studied species. Damin et al. observed losses of around 10 to 20% of applied N fertilizer after the application of glyphosate and glufosinate ammonium in Brachiaria decumbens and Pennicetum glaucum. In those same studies, a reduction of around 40% of the total N in the plant was observed after the application of herbicides. From a practical point of view, the effect of herbicides on the total N of plants used for soil covering is important because it can affect the mineralization rate due to a raise in the C:N ratio, and it can effect the immobilization of N by soil microorganisms, a process that reduces the nutrient's availability to plants.

N Mineralization/Immobilization

When organic residue is added to the soil, extra-cellular enzymes are liberated by the biota, which degrade the residue into substances with smaller molecular weights (monomers). These monomers are absorbed and metabolized by microbial cells, which transform them into inorganic forms, a process called mineralization. Mineralization occurs at the same time as immobilization, which refers to the incorporation of mineral nutrients to the microbial cells to meet the nutritional demand of the soil microbiota.

Table: Data of N recovered and lost from 15N-labeled fertilizer (Ndff) in the soil-plant system after herbicide application in Brachiaria decumbens (B.d.) and Pennicetum glaucum.

Treatments		Above ground part	Roots		Entire plant		Soil		Losses on soil-plant system	
	P. g.[1]	B. d.[2]	P. g.[1]	B. d.[2]	P. g.[1]	B. d.[2]	P. g.[1]	B. d.[2]	P. g.[1]	B. d.[2]
	g m-2									
Control	13.8 a	11.5	3.4 a	6.1 a	17.2 a	17.6 a	2.4 b	2.1 b	5.4 b	9.9 b
Glyphosate	10.8 b	8.1	2.7 b	2.9 b	13.6 b	11.1 b	3.4 a	3.4 a	8.0 a	15.1 a
Glufosinate	11.0 b	8.1	2.0 b	2.4 b	12.9 b	10.5 b	4.0 a	3.1 a	8.0 a	16.0 a
LDS (0,05)	2.1	3.8	0.5	1.8	3.0	4.9	0.7	0.8	2.3	4.9
P	**	ns	*	**	**	**	*	**	*	*
C.V. (%)	19.2	26.3	31.7	32.3	8.9	25.1	28.3	18.9	14.9	24.2

[1]Measurement of ten replicates. [2]Measurement of six replicates. *Means followed by different letters in columns differ from each other by Tukey's Test ($\alpha = 0.05$). **Means followed by different letters in columns differ from each other by Tukey's Test ($\alpha = 0.01$). The total dose of N was 25 g m^{-2} in pearl millet and 15 g m^{-2} in signal grass.

From the incorporation of vegetable residues in soil to their mineralization, several living organisms act in this process. In the initial phase, the vegetable residues are fragmented into smaller litter by the soil's macro-fauna. Subsequently, a diverse microbial community acts to liberate extra-cellular enzymes, which turn the substances that are easily decomposed into monomers that can then be absorbed by microbial cells. Substances that are difficult to degrade, such as cellulose and lignin, are degraded by a more specialized group of microorganisms, which compose the secondary and tertiary mineralizers. Residues that are rich in lignin and polyphenols have slower mineralization rates in the soil due to the resistance these substances have to degradation. The C/N ratio also has a major influence on the mineralization rate of vegetable residues, as it determines the N availability in the microbiota. Crop residues with a C/N ratio around 15:1 to 20:1 (e.g., soybeans, beans, lupine, forage turnips) tend to degrade faster than those with a C/N ratio higher than 30:1 (e.g., black oat, pearl millet, corn, sorghum). Heinzmann studied the N dynamics after winter green fertilization with black oat, forage turnip, hairy vetch, lupine and wheat and their effects on soy, corn and bean crops. He observed that the N liberation from legume and forage turnip residues occurred in the first weeks, whereas the black oat liberated more N in the blooming and graining stages.

Effect of Herbicides

In addition to chemical composition, the mineralization of crop residues added to the soil depends on factors that affect the composition and activity of the decomposing microbial community. These factors include temperature, humidity, aeration and the presence of toxic or nurturing substances. Concerning the effect of herbicides on mineralization rates, some observations should be made: 1) Herbicides can increase vegetable residue decomposition rates due to a greater detachment of leaves and roots caused by senescence. Thus, they may have a beneficial physical effect on stubble mineralization 2) However, the increased C/N ratio in the stubble caused by the application of these products may decrease the mineralization rate. 3) Furthermore, these herbicidal molecules may alter the soil's microbial community and cause an increase or a decrease in mineralization, depending on the affected microorganism populations. Argenta et al. observed an increase in the C:N ratio on the aerial parts of black oats 16 days after the application of glyphosate, glufosinate and paraquat. Although the paraquat did not decrease the N content of the plant during senescence, there was a remobilization of the nutrient to the reproductive structures and to the root system (Marschner, 1995). With the increase in the C/N ratio, the stubble mineralization rate can decrease. In fact, Damin et al. assessed the N decomposition rate present in pearl millet and black oat stubble after the application of glyphosate and glufosinate ammonium. They observed an increase in the stubble dry matter remaining on the soil surface and a decrease in N liquid mineralization. In their studies, the absorption of N in stubble (enriched with 15N) by corn plants was also assessed, and the herbicide-treated stubble was observed to be less effective in supplying N to plants. The mineralization of organic substances is mainly mediated by chemoorganotrophic microorganisms, which should not be damaged by the application of glyphosate and glufosinate ammonium because these herbicides act on metabolic routes that are typical of lithotrophic organisms (chemolithotrophic or photolithotrophic). Therefore, the reduction in mineralization observed by Damin et al. was attributed to the increase in the stubble C/N ratio.

In fact, Acinelli et al. observed that the application of the recommended doses of glyphosate and glufosinate ammonium did not interfere in the activity or in the soil microbial carbon biomass. The application of higher doses than recommended resulted in greater microbial activity, which can be connected to the death of lithotrophic micro-organisms, ensuring a competitive advantage to heterotrophs. Similar results were found for glyphosate by Haney et al. Ratcliffe et al. and Zabaloy & Gómez. Some studies have shown that the N mineralization that is already in the soil is increased by the application of glyphosate. Grossbard assessed the effect of glyphosate on soil N mineralization and observed an increase in mineralization when the herbicide was used. Haney et al. evaluated under laboratory conditions the effect of applying atrazine a mix of atrazine and glyphosate on the mineralization of total-C and total-N present in the soil. These authors observed a greater C and N mineralization in the soil treated with the mixture. The results showed that the glyphosate favored the chemoorganotrophic microbiota, causing an increase in the soil edaphic organic matter mineralization. In tillage systems in Brazil's Cerrado, a reduction in deposited stubble mineralization is desirable because it increases the permanence time of residue over the soil. However, the lower effectiveness of the residue in supplying N to plants and microorganisms can reduce productivity when the N supply through fertilization is not appropriate because the addition of low N content organic residues to the soil results in the immobilization of the nutrient by the microbiota, reducing its availability to plants. Therefore, the use of herbicides for covering crop desiccation in tillage system should be considered in the decision making processes of N fertilization management.

Table: Dry mass, carbon, nitrogen, and C:N ratio in above-ground parts of black oat treated with herbicides at 101-days-old grown in soil.

Treatments	DM	Carbon	Nitrogen	C:N	ΔDM	ΔC	ΔN
	------g per pot------		mg per pot		-------------------%------------------		
Control	6.5 b	2.6 b	117.2 b	22.6 b	49.9 a	49.4 a	69.6 a
Glyphosate	8.5 a	3.7 a	159.9 a	24.3 b	34.1 b	22.1 b	51.8 b
Glufosinate	9.3 a	3.8 a	160.2 a	28.6 a	28.3 b	23.2 b	49.7 b
F-test	**	**	**	**	**	**	**
CV (%)	15.5	13.6	16.2	12.4	25.6	30.3	12.3

DM = Dry mass; C = carbon; N = nitrogen; C:N = Carbon to Nitrogen ratio; Δ = (Non-recovered fraction/ total amount added)*100. Means followed by the same letter in a given column indicates nonsignificant differences at the 5% level by Tukey test. **, * = Significant at 1% and 5% levels, respectively.

N Exuding through the Roots

The rhizosphere may be defined as the soil zone that is influenced by the roots. The deposition of exuded matter and root fragments in this region favors the growth and activity of the soil microbial community, which has important functions in cycling and in the supply of nutrients to plants, especially N. About 15% of the total CO_2 set by plants is deposited in the rhizosphere in a cereal crop cycle. Sugars, amino acids, organic acids, phenols, exoenzymes and ions such as NH_4^+ are the main compounds exuded by the roots. Exudation may be an important way for N to exit the plant. Klein et al. estimated that the C:N ratio of exuded compounds for the Poaceae family species ranged from 2 to 2.7. However, other studies have demonstrated that the exuding of compounds containing N is very low, and the exuded compounds are rich in sugars. The main factors affecting the amount and composition of materialexuded from the roots are vegetable species, plant development stage, environmental stress, nutritional state, injuries, application of chemical products and activity and composition of the soil microbial community.

Table: Net N mineralization, non-recovered fraction and soil-plant recovery of the nitrogen arising from black oat residue with or without herbicide previous application.

Treatments	Net N mineralization	Non-recovered	Soil – Plant
	----------------------mg per pot-------------------------		
Control	201.2 a	52.9	418.0 a
Glyphosate	150.0 b	43.9	383.9 ab

Glufosinate	148.5 b	74.6	368.3 b
F-test	**	NS	*
CV (%)	8.8	54.6	8.9
		--%--	
Control	42.3 a	11.2	88.8
Glyphosate	35.3 b	16.8	83.1
Glufosinate	33.5 b	16.8	83.1
F-test	**	NS	NS
CV (%)	8.7	55.4	8.8

Means followed by the same letter in a given column indicates nonsignificant differences at the 5% level by Tukey's test; **, * = Significant at 1% and 5% levels, respectively; NS = nonsignificant difference.

The Effect of Herbicides in Exudated N

Herbicides may affect some of the mentioned factors and, consequently, root exudation. Kremer et al. evaluated amino acid and carbohydrate exudation after glyphosate was applied to conventional crops and crops genetically modified for herbicide tolerance. They concluded that the exudation was stronger in both crops after the product was applied. Similar results were obtained in bean plants that were desiccated with glyphosate. These authors found aspartic acid, glutamic acid, serine, glycine, amino butyric acid, valine, isoleucine, leucine, tyrosine, phenylalanine and proline in the material exuded from the roots. They observed that the valine, isoleucine and glycine concentrations were higher in the plants desiccated with the product.

The herbicides that were used as a replacement for the glyphosate also increased root exudation. Damin et al. evaluated the compound exudation containing N and the root detachment in Brachiaria decumbens plants after the application of glyphosate, glufosinate ammonium and paraquat. They observed an increase in exudation and the detachment of roots after the application of three products. Paraquat, unlike glyphosate and glufosinate ammonium, does not have its action connected to the N metabolism of the plant. As all herbicides increase N exudation, this process is probably not connected exclusively to the product's active mechanism. Physiologic events characteristic of senescence could justify the results obtained in this research. However, Kremer et al. observed that the exudation of sugars and amino acids was higher in soybeans genetically modified for tolerance to glyphosate than in the conventional soybeans. In this case, the increase in exudation was not connected to the development of senescence after the application of the herbicide.

It is worth stressing that the losses from exudation and root detachment together represented less than 4% of the N applied to the entire plant and, therefore, they should not affect in a relevant way the N content of the same plant. The compound N exudation, notwithstanding, may modify the composition and activity of the soil microbial community due to the supply of carbon sources that are highly soluble and readily available.

There is a great diversity of microorganisms in the soil, which, according to the kind of nutrition, may be grouped in photolithotrophic, photoorganotrophic, chemolithotrophic and chemoorganotrophic. Phototrophic microorganisms (photolithotrophic and photoorganotrophic) are rare, and chemotrophic microorganisms are abundant in the soil. All kinds of fungi are chemoorganotrophic, i.e., they use organic molecules as a source of carbon and energy, whereas bacteria and actinomycetes may come from a variety of groups. Processes such as nitrification and N biological fixation are mainly mediated by chemolithotrophic organisms, which use CO_2 as a carbon source and obtain energy from inorganic compound oxidation.

Due to the nutritional type, the supply of easily decomposable organic carbon sources caused by exudation offers a competitive advantage to chemoorganotrophic microorganisms, modifying the proportion of microorganism populations in the soil microbial community. In addition, chemoorganotrophs may be favored by the use of herbicides such as glyphosate and glufosinate, which act in metabolic routes that are present only in lithotrophic organisms (photolithotrophic and chemolithotrophic).

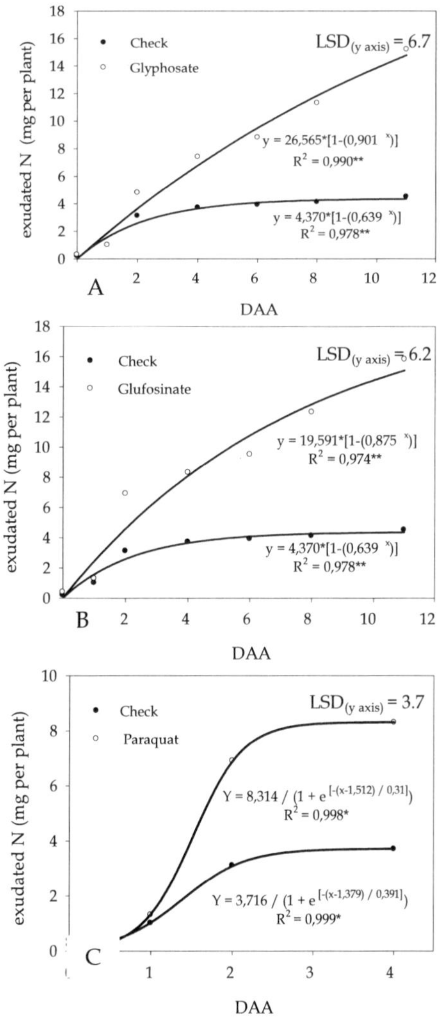

Time accumulated exudation of nitrogen after herbicide application on Brachiaria decumbens Stapf. A – Glyphosate (F_{herb}=2.8*/F_{DAA}=13.5**/$F_{Herb\ x\ DAA}$=4.3**); B – Glufosinateammonium (F_{herb}=4.9*/F_{DAA}=15.4**/$F_{Herb\ x\ DAA}$=4.9**); C – Paraquat F_{herb}=6.7*/F_{DAA}=15.2**/$F_{Herb\ x\ DAA}$=3.1*). DAA = days after herbicide application.

Among the chemoorganotrophic microorganisms that are important for agriculture are the decomposers of organic residues and most of the plants' pathogens. Kremer et al. observed an increase in fungi biomass of the kind Fusarium sp. after the application of glyphosate to soybeans. Liu et al. observed an increase in the number of colonies and a higher development of the Pytium sp. fungi after addition to the environment through the bean roots cultivated with glyphosate desiccated or exuded herbicide. As the glyphosate can also be exuded by the roots, it is possible that the population increase of pathogenic microorganisms after the application of the herbicide is connected to the N compound exudation as well as to the presence of the herbicide.

Aside from those processes, a higher colonization of roots by pathogenic microorganisms may be associated with the effects the herbicide has on the plant's defense system. Liu et al. also observed that bean plantlets growing in the middle of Pytium sp. showed a higher lignin content when glyphosate was not added to the environment. Studies in sterile environments have shown that the increase in Pythium and Fusarium sp. is connected to the effectiveness of the glyphosate herbicide.

It is likely that other chemoorganotrophic microorganisms, such as denitrificants, are favored by the use of glyphosate and glufosinate, as the highest exudation results in increased C and N availability in both the soil and these herbicidal molecules (unlike paraquat) do not affect the soil's chemoorganotrophic microbiota.

Nitrification/Denitrification

In addition to reducing the agronomic effectiveness of N fertilization, N losses because of the denitrification-nitrification processes in soil may have a negative impact on the environment because considerable amounts of N_2O are generated in these processes. This gas is considered the fourth-greatest contributor to the greenhouse effect, although it is present in the atmosphere and in the stratosphere in small amounts. Moreover, N_2O is an intermediary in the ozone layer destruction process. It is estimated that approximately 20% of the global gas emissions of nitrogen oxides and elemental N are connected to agricultural activity. However, in countries such as Australia, agriculture may be responsible for 80% of these emissions. The main processes that result in the gas loss of N oxides and elemental N are nitrification and denitrification. During nitrification (in the oxidation of NH_4 into N_2O), there may be slight production of nitrous oxide by a chemical dismutation of nitroxyl (NOH) or by the action of nitrite reductase. In this case, the formation of N_2O may occur under aerobic conditions, during the nitrification. Although the N_2O production by nitrification might, in some situations, have an emission potential similar to that of the denitrification process, denitrification seems to be the main N_2O source in the soil. Biological denitrification is defined as an anaerobic breathing process mediated by microorganisms that are able to use NO_3 or NO_2 as final electron acceptors. The main products of denitrification are N_2 and N_2O, and the proportion of production of each gas varies according to the environmental conditions. There are more than 125 bacterial species that are capable of performing denitrification, including phototrophs, lithotrophs and organotrophs (the most important group). The main environmental factors that interfere in denitrification rates are O_2 supply, water content, temperature, organic matter, the presence of organic substances and ammonium (NH_4^+) and nitrate (NO_3^-) concentration. A number of studies have demonstrated that an increase in carbon content results in higher N_2 and N_2O emissions. Furthermore, pH may interfere in N_2O and N_2 emissions, as it affects nitrification, when it is reduced 3.5.

The Effect of Herbicides in Nitrification/Denitrification

Chemoorganotrophic microorganisms may be favored by the application of some herbicides, and such is the case for glyphosate and glufosinate ammonium. Laboratory studies have demonstrated the increase of N_2O emission in soils treated with glyphosate, which might be attributable to the herbicide serving as a nutrient source to organisms and offering a competitive advantage to denitrificants through the killing of chemolithotrophic microorganisms. However, higher N_2O emissions were also observed for soils with plants desiccated by these molecules. This observation may be attributed to the indirect effects of the herbicide in the rizosphere due to the killing of the plants, such as the increase in promptly available carbon content and a reduction in the environment's NO_3 absorption, leaving it more susceptible to microbial attack.

Most nitrifying microorganisms are chemolithotrophic and, therefore, they can show the metabolic routes present in the plants, including those that were affected by glyphosate or glufosinate ammonium. Damin et al. evaluated the NO_3^- content of two soils (haplustox and quartpsament) 90 days after the permanence of millet stubble desiccated with these herbicides. The N-NO_3^- content in the glyphosate and glufosinate treatments was reduced in the quartpsament, but not in the haplustox. The haplustox soil had a higher sorting capacity than the quatzpsament, as it could be noticed in the contents of the clay, the CTC and the organic matter of soils. In fact, the NH_4^+ availability to nitrifying microorganisms was lower in the haplustox than in the quatzpsament due to the higher ion absorption to the colloidal fraction of the soil. Moreover, the glyphosate showed a great affinity toward the iron and aluminum oxides and the hydroxides common in the oxisoils. The strong connection between the herbicide and these fractions reduced the bioavailability of the product to the microorganisms.

Some authors have observed a rise in NO_3^- content in the soil after the plants are killed. This result can be attributed to a higher availability of NH_4^+ in the soil as it stops being absorbed by the plant. Although the decrease in nitrification indicates a damaging effect to the microbiota, it is important to highlight that the conversion of NH_4^+ into NO_3^- in the soil may promote a higher risk of superficial water contamination by nitrate and a lower availability of the nutrient to future crops.

Table: NH_4^+ and NO_3^- content in two soils 90 days after the permanence on the soil of Pennicetum glaucum stubble desiccated with herbicides.

	NH_4^+				NO_3^-		
				mg per pot			
Treatments	Quartpsament		Haplustox		Quartpsament		Haplustox
Witness+	229.7		297.7		117.2baA		31.5 aB
Glyphosate	197.7		331.6		59.3 bA		34.2 aA
Glufosinate	212.6		314.5		82.6 bA		21.4 aB
Average	213.3 b		314.6 a		86.4		29.1
Herbicide F		0.0 ns				3.8*	
Soil F		24.3**				43.7**	
Interaction F		0.9 ns				4.1*	

+Witness (stubble with no application of herbicides); Means followed by the same letter in a given column indicates nonsignificant differences at the 5% level by Tukey's test. **, * = Significant at 1% and 5% levels, respectively.

Biological N fixation

Biological N_2 fixation (BNF) is an important entrance path for N in agroecosystems, and it is the main source of the nutrient in leguminous farming. In these plants, BNF is performed through the symbiosis between N fixing bacteria (NFB), which break the N_2 triple bond through the action of the nitrogenase enzyme, reducing it to NH_3, while the plants supply photosynthates to the bacteria.

Only a small percentage of prokaryotes are able to accomplish BNF. Among these prokaryotes, however, there is high morphological, physiological and philogenetic diversity. In addition, N fixing micro-organisms are found among the heterotrophs, the anoxygenic phototrophs, the archaebacteria and the cyanobacteria. The main kinds of fixing bacteria, which are economically important, are Rhizobium and Bradyrhizobium, which are predominantly chemoorganotrophic metabolic proteobacteria, although there are some chemolitotrophic strains of Bradyrhizobium japonicum.

Effect of Herbicides

Due to the diverse kinds of metabolism of N fixing bacteria, it is difficult to establish which conditions are favorable or unfavorable to NFB, as the effects of a determined factor will depend on the bacteria community present in the soil. This difficulty may be one of the reasons for the contradictory results in the literature concerning the effect of herbicides on nodulation. The growth, survival and nitrogenase activity results vary according to the herbicide doses applied, the species/strains studied and the experimental conditions. Recently, Santos et al. observed differentiated responses in the growth of strains of native Brazilian Bradyrhizobium spp. because of the application of different commercial formulations of glyphosate. This differentiation reveals the importance of considering the adjuvant components of those herbicides in impact evaluations. With soybeans that are tolerant to glyphosate reported that the herbicide increased the number and decreased the weight of nodes. Jaworski observed a reduction of BNF after the application of glyphosate. Some authors have attributed the BNF reduction after glyphosate application to the inhibition of the EPSPs enzymes in microorganisms. Moorman evaluated the effect of glyphosate on the development of Bradyrhizobium spp. in the environment with and without the addition of amino acids. It was observed that the herbicide reduced the Bradyrhizobium spp. population only when no amino acids were added. It is worth stressing, however, that only chemolithotrophic microorganisms, such as some strains of Bradyrhizobium spp., should show this metabolic route. Some authors have attributed the BNF reduction after glyphosate application to the physiological changes caused to the plants. It is common for nutritional disorders to happen in transgenic soybeans after glyphosate application, and two of them are particularly important: manganese and iron. The lack of manganese in the tissue may increase the ureide concentration in the aerial parts, signaling to the BNF to stop fixation. Moreover, the application of the herbicide reduces the production of flavonoids, substances that stimulate the genetic expression of nodulation.

In conclusion, the effect of herbicides on a soil-plant system's N loss and on the mineralization rates of vegetable residues should be considered when predicting the N availability to plants, in nutrient-caused environmental impact studies, in production cost estimations and in decision making concerning the appropriate time to apply N doses in areas under a tillage system. This kind of information is important for creating managing strategies that increase the efficiency of N fertilization and reduce the environmental impacts caused by the use of fertilizers and herbicides.

Effects of Herbicides on Aquatic Ecosystems

Offsite Movement and Direct Applications

The use of herbicides and other chemical agents in agriculture may result in accidental introduction into waters. When it is applied as post emergence spray, herbicides may enter aquatic systems through accidental offsite movement in herbicide spray drift, or through transport in leaching and surface run-off. Particularly, under field conditions, glyphosate is usually assumed to be rapidly and tightly adsorbed to soil. Consequently, glyphosate is unlike to leach into ground waters or run-off significantly into surface waters following application. In several laboratories and some field studies, the immobility of glyphosate in soils has been demonstrated. In contrast, other field studies showed detectable concentrations of glyphosate in flume and streams after applications. Even thought it was only restricted to a relatively brief window of time (about 1 day post application), due to the fast dissipation kinetics of glyphosate in the field. In natural waters, glyphosate dissipate rapidly (half lives < 4 days) being removed from water due to adsorption to suspended particulates followed by subsequent sedimentation and or biodegradation. However, longer half lives were reported in hard waters, where glyphosate residues could be measured after 11 days post application. Edwards et al. reported important glyphosate concentrations in runoff from natural rainfall following early springtime treatments in no-tillage agriculture soils. The highest concentration of glyphosate in runoff waters (5.2 mg/L) was found in runoff occurring 1 day after treatment at the highest rate (8.6 Kg/Ha of Roundup). The maximum amount of glyphosate transport by runoff was 1.85% of the amount applied, most of which occurred during a single storm on the day after application. In addition, Feng et al. found in a treated watershed, a dramatic increase of glyphosate concentrations (about 1.1 and 1.5 mg/L) in two oversprayed streams in response to first rainfall event 27 h post application. Authors attributed their observations to several source of input mobilization of residues in ephemeral stream channels feeding the tributary; wash off of unabsorbed residues from overhanging vegetation, surface runoff and subsurface flow.

Regarding POEA residues, based on adsorption and degradation data, leaching and runoff potential is expected to be small. POEA strongly adsorb to soil, although little information about POEA offsite movement is nowadays available. Offsite movement of glyphosate is also possible through spray drift. Although the spray drift of pesticides is not compound specific, this is relevant when non-target effects of glyphosate based herbicides are considered, and several studies have specifically addressed the issue. Some studies reported that the spray deposition decreased to around 10 % of the application rate in the first 30 m and less to 5 % at a distance of 200 m. Other studies suggested that drift rates would be greater. For instance, residues have been measured 400 m downwind from applications sites. Considering offsite movement of glyphosate from treated soils through drift and run-off, Giesy et al. estimated an acute scenario considering worst-case exposure conditions. The estimate was based on two assumptions, (a) that runoff (2%) from 10 Ha field treated at the maximum single use rate of Roundup entered to 1 Ha pond (2 m deep) and (b) that 10% of maximum single application rate per hectare entered the pond through drift, assuming aerial application. Based on these assumptions, maximum concentrations of Roundup in natural water would range from 0.27 to 0.41 mg/L.

However, clearly higher concentrations in surface waters could be expected if assumptions are changed. For instance, some authors have reported that glyphosate can be readily desorbed from soil and has the potential to be extensively mobile in the soil environment. Adsorption of glyphosate to soil particulates is determined by chemical and physical characteristic of soils, which in turn affect the potential for off-target movement of the herbicide through water runoff or subsurface flow. Interestingly, given that glyphosate is bound to soil through its phosphonic acid moiety, the addition of inorganic phosphorus could potentially release glyphosate from soil particles through competition for sorption sites. Piccolo et al. reported in an experimental study with some European soils that desorption varied from around 15 to 80% of the absorbed herbicide according to the soil characteristic. These observations, as well as supposing higher rates of terrestrial uses and higher spray drift due to weather conditions, could elicit elevated off-target movements of glyphosate formulations in to water ecosystems. Particularly, these impacts will be more important in ponds, ephemeral streams and ditchbank areas of irrigation canals due to their low water volume, and higher perimeter and area /volume proportions. On the other hand, some glyphosate based herbicides were specially formulated to be used as aquatic herbicides, and have been employed extensively to control noxious aquatic weeds and algal blooms. For this purpose, glyphosate based herbicides are directly applied in aquatic ecosystems and their residues can be expected to be higher than that resulting from agricultural and other non aquatic uses.

Furthermore, glyphosate can move considerable distances in canal or stream waters affecting undesired places. Fifty-eight percent of applied glyphosate was detected at distances 8 and 14.4 Km downstream from sites of introduction. Regardless herbicide sources, it is very important to set up the amount of glyphosate that have been measured in the field. Unluckily, there are few relevant field data on the concentration of glyphosate in aquatic habitats. The highest concentrations that have been observed in nature were: 1.24 mg a.e./L; 1.54 mg a.e./L; 2.8 mg a.e./L and 5.2 mg a.e. /L.

Toxicity of Glyphosate based Herbicides in Aquatic Environments

Toxicity Assessment

Here,we will study the contributions about glyphosate, glyphosate formulations and surfactants effects on non-target aquatic organisms. Different parameters (lethal and sublethal effects) were evaluated in reviewed studies to characterize the hazard of chemicals (e.g. mortality, growth, biomass, 14C uptake, weight, density, length, pigments, mobility, reproduction, metabolism, etc). Results were expressed as LC (concentration lethal to 10% and 50% of test organisms), EC (effective concentration causing specified effects in 10% and 50% of test organisms) and IC (inhibition concentration to specified effects in 10% and 50% of test organisms). In addition, when available, values of NOEC (no observed effect concentration) and LOEC (lowest observed effect concentration) were pointed out. When dose-response curves were not available or were not calculated due to experimental design (e.g. field studies carried out in micro- and mesocosms), contributions were described as concentrations of treatments and obtained outcomes (generally in % of control values). A complete resume of published outcomes in acute and chronic tests and field experiments were listed in tables. Concentrations were preferentially expressed as were originally reported in each reviewed contribution. In order to relate the aquatic toxicity of the herbicide to realistic exposure levels, the expected environmental concentration (EEC) was taken as a reference value. We considered a EEC of 2.6 mg a.i./L,. Similar values were estimated by other authors: 1.87 mg a.i./L and 3.73 mg a.i./L; though higher values were also evaluated.

Effects on Non-target Aquatic Plants and Algae

Herbicides are manly designed to kill unwanted terrestrial plants. Consequently the most sensitive group of aquatic non-target organisms is expected to be aquatic plants and algae. Aquatic plants and algae play a pivotal role for the function of aquatic ecosystems. Aquatic plants aid in stabilizing the sediment both in lakes and running waters, and their presence affects sedimentation rates, flow velocity, nutrient uptake and recirculation. In addition, they provide refuges for insects, crustaceans and fish, and act as substrates for surface-living microorganisms, snails and other epiphyte grazers. Microalgae (phytoplankton and periphyton communities) provide the basis for a range of food-webs in the aquatic environment and are therefore fundamental to the functioning of aquatic ecosystems. Single species test in algae and aquatic plants treated with glyphosate alone (i.e. technical grade acid or salts of glyphosate), showed a wide range of EC and IC values; indicating different sensibilities. Microalgae presented EC_{50} values for glyphosate treatments ranging from 0.68 mg a.e./L in the diatom Skeletonema costatum to around 600 mg a.e./L in the green algae Chlorella pyrenoidosa. It is important to clarify that these values indicate the concentration that elicited the 50 % of reduction in the evaluated parameter. Some works showed that 10 % of reduction (EC_{10}) could be reached between 3 to 16 folds lower concentrations than EC_{50}. For instance, 10% growth inhibition in the green algae Scenedesmus subpicatus was observed in treatments with 1.6 mg/L of glyphosate acid.

In addition, Christy et al. reported a 10% growth inhibition in Chlorella sorokiniana at the concentration of 2 mg a.e./L. Regarding macrophytes, generally lower values of EC and IC were reported, indicating a higher sensibility. For example, IC_{50} and EC_{50} values ranged from 0.22 mg a.i./L for Myriophyllum aquaticum to 46.9 mg/L for Lemma minor. The relative toxicity of glyphosate itself vs. commercial formulations and surfactants only can be evaluated in studies specially designed to this purpose. In general, commercial formulations were more toxic than glyphosate alone. For example, Tsui & Chu observed a 7 folds higher toxicity of Roundup than the IPA salt of glyphosate in the green algae Selenastrum capricornutum. Alike results were reported for Selenastrum capricornutum and the macrophyte Lemma minor, showing 4 folds higher toxicity of Roundup than glyphosate. Lower differences in toxicity were registered by other authors, reporting between 1.2 to 1.8 folds higher toxicity of commercial formulations than active ingredient. POEA itself contributed to Roundup toxicity with values ranged from about 45% for Skeletonema costatum to 85 % for Selenastrum capricornutum. Numerous studies have been published about pesticide toxicity assessment on microalgae, using single species tests.

However, Bérard et al. demonstrated that single-species tests may fail to predict indirect or system responses to toxicants, such as changes in population competition or succession. According to these authors, studies focusing on the whole natural community provide more reliable predictions about herbicide safety in aquatic environments. In studies assessing communities, significant direct and indirect effects of commercial glyphosate formulations have been reported. For example, Schaffer & Sebetich reported an increment of 161 % in net primary production for phytoplankton community treated with 0.13 mg a.i./L of Rodeo (commercial formulation without POEA). In contrast, Goldsborough & Brown, registered a 50% of reduction in periphyton primary production at values varying from 35.4 to 69.7 mg a.i./L of Roundup. However, in this contribution, 4 of the 6 treated ponds showed a reduction in the mean values of primary production with much lower concentrations (a dosage of 0.89 mg a.i./L). In microcosms experiments with natural marine microbial community, significant effects in species number and relative abundance of phytoplankton were observed at 10 μg a.i./L of Roundup. Comparable results were obtained by Pesce et al., reporting changes in riverine algal communities exposed to about 10 μg/L of glyphosate alone, in a microcosms experiment.

In addition, mesocosms studies showed remarkable results with a single pulse application of Roundup at concentrations of 6 and 8 mg a.i./L. Even if these two contributions assessed herbicide effects in worst case scenarios, glyphosate concentration at the end of the experiments (11 and 14 days respectively), were around 2 mg a.i./L and effects were still clearly observed. At day 11, significant differences were observed in chemical and biological variables. For example, we observed changes in phytoplankton assemblage fractions, with a reduction of micro and nanophytoplankton densities (2.5 folds) and a concomitant increase of picocyanobacteria (PICY) densities (40 folds) accompanied by an increase of primary production. These results can be expected by either direct effect of herbicide (differences in sensibility among species) or indirect effects duo to interspecific competition. In addition, Vera et al. found that Roundup produced a clear delay in periphytic colonization in treated mesocosms. The periphytic mass variables: dry weight (DW), ash-free dry weight (AFDW) and chlorophyll a, were always higher in control mesocosms. Despite the mortality of algae, (mainly diatoms), cyanobacteria was favoured in treated mesocosms. We also observed that Roundup produced a long term shift in the typology of mesocosms, "clear" turning to "turbid" state due to an eutrophication process.

Table: Effects of glyphosate, different commercial formulations of glyphosate and POEA on algae and aquatic plants.

AQUATIC ALGAE & PLANTS	ASSESSED CHEMICAL	STUDY TYPE	ASSESSED PARAMETER	EFFECTS CONCENTRATION (mg/L) ##
- Phytoplankton and Periphyton				
Chlorococcumhypnosporum	Gly. (acid)	SST (7 d.)	Growth	96h EC_{50} = **68.0**
Chlorella pyrenoidosa				96h EC_{50} = **590.0**
Skeletonema costatum	Gly. (acid)	SST (7 d.)	Biomass	EC_{50} = **0.64**; NOEC = **0.28**
(Periphyton community)	Roundup®	LTNC (4 h.)	NPP	4h EC_{50} = (between **35.4** to **69.7**)
Scenedesmus acutus	Gly. (IPA salt)	SST (4d.)	Density	96h EC_{50} =**10.2**; NOEC = **2.0**
	Ron-do®			96h EC_{50} = **9**; NOEC = **3.2**
Ankistrodesmus sp	Rodeo®	SST (10 d).	Density	96h EC_{50} = **74.0**
(Phytoplankton community)	Rodeo®	LTNC (6 h.)	NPP	**0.13** mg/L elicited NPP increment
Selenastrum capricornutum	Gly. (IPA salt)	SST (4 d.)	Growth	96h EC_{50} =**41.0**
	Roundup®			96h EC_{50} =**5.81**
	POEA			96h EC_{50} =**3.91**
Skeletonema costatum	Gly. (IPA salt)	SST (4 d.)	Growth	96h EC_{50} =**5.89**
	Roundup®			96h EC_{50} = **1.85**
	POEA			96h EC_{50} =**3.35**
Selenastrum capricornutum	Gly. (n.c.)	SST (2 d.)	Growth	48h EC_{10} = **95.5**; 48h EC_{50} = **270.0**
	Roundup®			48h EC_{10} = **13.6**; 48h EC_{50} = **64.7**

(Phytoplankton & periphyton community)	Roundup®	MES (11 d.)	Density #	**6** mg/L elicited changes in community structure
(Microbial community)	Roundup®	MIS (7 d.)	Community structure #	**10** µg/L elicited a reduction in species number.
(Microbial community)	Gly. (n.c.)	MIS (14 d.)	Community structure #	~ **10** µg/L elicited changes in algal community structure.
Chlorella sacchorophila	Gly. (acid)	SST (3 d.)	Growth	72h EC_{10} = **3.0**;72h EC_{50} = **46.6**
Scenedesmus subspicatus				72h EC_{10} = **1.6**; 72h EC_{50} = **26.0**
(Periphyton community)	Roundup®	MES (42 d.)	Density #	**8** mg/L elicited changes in community structure
- Macrophytes				
Myriophyllum sibiricum	Gly. (acid)	SST (14d.)	Root length #	14d IC_{10} = **0.59**;14d IC_{50} = **0.84**
	Roundup®			14d IC_{50} = **1.22**
Myriophyllum aquaticum	Gly. (n.c.)	SST (14d.)	Growth & chl *a* #	14d EC_{50} = **0.22** (for growth) 14d EC_{50} = **0.22** (for chl a)
Lemma minor	Gly. (n.c.)	SST (7 d.)	Growth	7d EC_{10} =**3.8**; 7d EC_{50} = **46.9**
	Roundup®			7d EC_{10} =**3.5**; 7d EC_{50} = **11.2**
Myriophyllum spicatum	Roundup®	SST (21 d.).	Weight & length #	21d IC_{50} = **1.0** (for weight) 21d IC_{50} = **2.8** (for length)
Lemma gibba	Gly. (acid)	SST (10 d.).	Growth #	10d IC_{10} =**4.6**; 10d IC_{50} = **20.5**
	Roundup®			10d IC_{10} =**2.5**; 10d IC_{50} = **11.6**

Abbreviations and acronyms: gly. (glyphosate); n.c. (not clarified); n.a (not available); d. (days); SST (single species laboratory tests); LTNC (laboratory tests with natural communities), MES (mesocosms studies); (MIS) microcosms studies; SSFE (single species field experiments); chl a (chlorophyll a); NPP (net primary production). Notes: (##) Effects concentrations were expressed as mg/L of formulation, mg a.i./L or mg a.e./L, see bibliographic references to clarify. (#) Several parameters were assessed in these contributions; remarkable examples of the reported outcomes were listed in tables.

Effects on Non-target Aquatic Bacteria and Protozoa

The majority of the available pesticide data regarding aquatic microorganisms is for algae. Far fewer pesticide studies exist for aquatic bacteria and protozoa. Aquatic bacteria and protozoa (e.g. amoeboids, flagellates, ciliates and sporozoans) have key roles in the functioning of aquatics environments. Shortly aquatic bacteria occupy an important position in the aquatic food web since they are major actors in the decomposition of dead material, and thereby in the recycling of nutrients and carbon. They are extremely important in "Lake metabolism", being involved in mineralization processes and in the chemical transformation of elements between reduced and oxidized forms Protozoans are ecologically important as key links in food chains. Ubiquitous in aquatic environments, protozoans prey upon algae, bacteria, and other organisms and are themselves consumed by animals such as microinvertebrates. Thus, the ecological role of protozoa in the transfer of bacterial and algal production to successive trophic levels is very important (in the traditional food web and in the microbial loop).

On the other hand, some protozoa are important as parasites and symbionts of multicellular animals. Concentration effects of glyphosate itself on bacteria and protozoa varied widely and seem to indicate low sensibility. For instance, EC_{50} values obtained in treatments with glyphosate ranging from 18.2 mg/L for the bacteria Vibrio fischeri to 386.0 mg a.e./L for the ciliate Tetrahymena pyriformis. However, lower concentrations have been reported to produce observable effects. These authors registered a LOEC value of 5 mg/L for the parasite ciliate Ichthyophthirius multifiliis treated with glyphosate acid. Roundup showed higher toxicity than glyphosate for bacteria and protozoa in the revised bibliography. Tsui & Chu reported 6 folds higher sensibility of Vibrio fischeri to Effects of glyphosate, different commercial formulations of glyphosate, AMPA and POEA on bacteria and protozoa Roundup than to glyphosate acid. In addition, these authors observed 2.7 and 13 folds higher sensibility of the ciliates Euplotes vannus and Tetrahymena pyriformis to Roundup,respectively. In addition, the ciliate parasite Ichthyophthirius multifiliis showed an accentuated response, being several times more sensible to Roundup. Values of EC_{50} obtained in Roundup treatments ranged from 23.5 to 29.5 mg a.e./L; though lower values produced observable effects. For example, Everett & Dickerson registered LOEC values of 0.07 and 0.31 mg a.e./L for two ciliates. Besides, values of 0.1 mg a.i./L of Roundup and Avans (other glyphosate commercial formulation) elicited reduction of at least 50% in the swimming velocity of Euglena gracilis.

Table: Effects of glyphosate, different commercial formulations of glyphosate, AMPA and POEA on bacteria and protozoa.

MICROORGANISMS	ASSESSED CHEMICAL	TYPE OF STUDY	ASSESSED PARAMETER	EFFECTS CONCENTRATION (mg/L) ##
-Bacteria and Protozoa				
Ichthyophthirius multifiliis	Gly. (acid)	SST (5 h)	Mortality	NOEC = **2.5**, LOEC = **5.1**
	Roundup®			NOEC = n.a., LOEC = **0.07**
Tetrahymena thermophila	Gly. (acid)	SST (24 h)	Mortality	NOEC = **10.1**, LOEC = n.a.
	Roundup®			NOEC = n. a., LOEC = **0.31**
Brachionus calyciflorus	Gly. (n.c.)	SST (24 h)	Growth #	24h EC_{50} =**28.0**
Vibrio fischeri	Gly. (IPA salt)	SST (5 min.)	Growth	5min EC_{50} =**162.0**
	Roundup®			5min EC_{50} =**24.9**
	POEA			5min EC_{50} =**10.2**
Tetrahymena pyriformis	Gly. (IPA salt)	SST (40 h.)	Growth	40h EC_{50} =**386.0**
	Roundup®			40h EC_{50} =**29.5**
	POEA			40h EC_{50} =**4.96**
Euplotes vannus	Gly. (IPA salt)	SST (48 h.)	Growth	48h EC_{50} =**64.1**
	Roundup®			48h EC_{50} =**23.5**
	POEA			48h EC_{50} =**5.0**
Euglena gracilis *	Roundup®	SST (7 d.)	Velocity #	NOEC = **0.05**, LOEC = **0.1**
	Avans®			NOEC = **0.05**, LOEC = **0.1**
Vibrio fischeri	Gly. (acid)	SST (15 min.)	Bioluminescence	15min EC_{50} =**18.2**
	AMPA			15min EC_{50} =**53.4**
Tetrahymena pyriformis	Gly. (acid)	SST (45 min.)	Enzyme activities #	45min EC_{50} =**87.9**
	AMPA			45min EC_{50} =**166.5**

Notes (*) Euglena gracilis is a mixotrophic green flagellated; although in this resume was grouped with protozoa.

Table: Effects of glyphosate, different commercial formulations of glyphosate and POEA on aquatic invertebrates.

AQUATIC INVERTEBRATES	ASSESSED CHEMICAL	STUDY TYPE	ASSESSED PARAMETER	EFFECTS CONCENTRATION (m/L) ##
-Crustaceans (copepods, cladocerans and amphipods)				
Daphnia magna	Roundup®	SST (48h)	Immobilization	48h EC_{50} = **3.0**
Gammarus pseudolimnaeus	Roundup®	SST (48h)	Mortality	48h LC_{50} = **62.0**
Daphnia pulex	Roundup®	SST (96 h.)	Immobilization	96h EC_{50} = **8.5**
	POEA			96h EC_{50} =**2.0**
Daphnia magna	Ron-Do®	SST (48h)	Immobilization	48h EC_{50} = **61.7**
Daphnia spinulata	Ron-Do®	SST (48h)	Immobilization	48h EC_{50} = **66.2**
Simocephalus vetulus	Vision®	SST (8 d)	Survival and reproduction #	**0.75** mg/L elicited survival and reproduction reduction
Ceriodaphnia dubia	Rodeo®	SST (48 h.)	Mortality#	48h LC_{50} = **415.0**
	Roundup bioactive®			48h LC_{50} = **81.5**
	Roundup®			48h LC_{50} = **5.7**
Hyalella azteca	Rodeo®	SST (48 h.)	Mortality#	48h LC_{50} = **225.0**
	Roundup bioactive®			48h LC_{50} = **120.0**
	Roundup®			48h LC_{50} = **1.5**
Ceriodaphnia dubia	Gly. (IPA salt)	SST (48 h.)	Mortality	48h LC_{50} = **415.0**
	Roundup®			48h LC_{50} = **5.4**
	POEA			48h LC_{50} = **1.2**
Acartia tonsa	Gly. (IPA salt)	SST (48 h.)	Mortality	48h LC_{50} = **49.3**
	Roundup®			48h LC_{50} = **1.77**
	POEA			48h LC_{50} = **0.57**
Daphnia magna	POEA (5:1)	SST (48 h.)	Mortality	48h LC_{50} = **0.18**
	POEA (10:1)			48h LC_{50} = **0.097**
	POEA (15:1)			48h LC_{50} = **0.85**
-Molluscs (snails and mussels)				
Pseudosuccinea columella	Gly. (n. c.)	SST (12 d.)	Growth and hatching #	**1** mg/L elicited growth increment and **10** mg/L inhibited hatching.
Pseudosuccinea columella	Gly. (n. c.)	SST (12 d.)	Metabolism	**0.1** mg/L elicited an increment in free amino acids.
Utterbackia imbecillis	Roundup®	SST (24 h.)	Mortality #	48h LC_{50} = **18.3**
-Others (insects and worms)				
Chironomous plumosus	Gly. (acid)	SST (48h)	Immobilization	48h EC_{50} = **55.0**
	Roundup®			48h EC_{50} = **18.0**
	POEA			48h EC_{50} = **13.0**
Lumbriculus variegatus	Gly. (acid)	SST (2-4 d.)	Bioaccumulation & enzyme action	**0.05** mg/L elicited increase in sGST activity.
	Roundup Ultra®			**0.05** mg/L elicited an increase in sGST activity.

Surfactant POEA itself resulted more toxic to bacteria and ciliates (between 2.4 to 5.8 folds) than Roundup; though the degradation product of glyphosate (AMPA) resulted less toxic than glyphosate acid.

Effects on Non-target Aquatic Invertebrates

Invertebrates comprise a large group of aquatic species with a wide variety in shape and size, and evolved to utilize different habitats and resources (e.g., insects, worms, snails, hydroids, crustacean, etc.). They can be practically divided in micro and macro groups. Microinvertebrates (zooplankton) are keystone species (e.g. as food for predators and top-down controller of phytoplankton, periphyton and detritus). The microinvertebrates (rotifers and copepods) are usually more abundant in number of individuals, but their smaller size limit their impact to size discrimination of phytoplankton rather than a reduction of total algal biomass. Large cladocerans (e.g., Daphnia spp.) can feed efficiently on a wide range of particle types and sizes. Moreover, cladocerans are also important as a major food source for many fish species and predatory invertebrates. Some crustacean species (e.g., shrimps, crabs and crayfish) are important as food resource for men. Insects are foraging on zooplankton, periphytic algae, and detritus and themselves prey for fish and waterfowl. In addition, aquatic molluscs (i.e. snails and mussels) have also a significant importance in aquatic environments and in human life (e.g. parasite vectors; invasive species; top-down controller of phytoplankton, periphyton and detritus; source of food to higher trophic levels, as well as human food resources).

Aquatic invertebrates seem to have low sensibility to glyphosate itself. Values of LC_{50} obtained in treatments with glyphosate ranged from 49.3 mg a.e./L for the marine copepod Acartia tonsa to 415 mg a.e./L for the cladoceran Ceriodaphnia dubia. However, Tate et al., reported remarkable results in the snail Pseudosuccinea columella (an intermediate snail host of Fasciola hepatica) treated with lower glyphosate concentrations (0.1, 1 and 10 mg/L). Concentrations of 1 mg/L elicited an increment in the growth in the third-generation of snails, as well as 0.1 and 10 mg/L elicited an inhibition of eggs hatching, abnormalities and polyembryony. Same authors observed significant differences in the metabolism of P. columella. Glyphosate concentrations of 0.1 mg/L induced about 2 folds increment in five free amino acids. In addition, Contardo-Jara et al. reported significant increment in the enzymes activities (sGST and SOD) of worm Lumbriculus variegatus at 0.05 and 0.1 mg a.i./L of glyphosate.

On the other hand, invertebrates showed higher sensibility to commercial formulations. For instance, Roundup showed between 3 folds to 76 folds higher toxicity than glyphosate itself. Values of LC_{50} obtained in Roundup treatments ranged from 1.5 mg a.e./L for the amphipod Hyalella azteca to 62.0 mg a.i./L for other amphipod Gammarus pseudolimnaeu. In addition, 0.7 mg a.e./L of Vision (a commercial formulation containing POEA) elicited a 100 % of mortality and more than 50 % reduction in total neonates per female in the cladoceran Simocephalus vetulus at values of Ph = 7.5. Sublethal effects were observed at much lower concentration of Roundup (1.1 μg/L) in the Clam, Ruditapes decussates, showing histological alterations. The surfactant POEA was several times more toxic for invertebrates than glyphosate itself and Roundup. For example, Folmar et al. reported almost 1.4 folds higher toxicity of POEA relative to Roundup in the midge larvae Chironomous plumosus; contributing with about 66 % of the Roundup toxicity. Higher values were reported by Servizi et al. and Tsui & Chu, being POEA 3 and 5 folds more toxic than Roundup, respectively. POEA can contribute with more than the 90% of Roundup toxicity On the other hand, commercial formulations without POEA (e.g. Ron-Do, Rodeo and Roundup bioactive) showed lower toxicity than other formulations.

Effects on Non-target Aquatic Vertebrates

Fishes are well appreciated in human societies in many ways (e.g., economical, recreational, ecological). In many countries, commercial fishing has a large economic importance as national food supply and as an export product. Fishes are one of the most demanded pets and there are a lot of people who enjoy sport fishing. This vertebrate group has an amazing diversity in morphology, size and color, which reflects their life history adaptations (e.g., feeding behavior reproduction and habitat selection). It is well known that fish populations have both direct and indirect effects on ecosystem function and structure in general (e.g., nutrient dynamics and cycling, zooplanktonic community composition), and especially in freshwater ecosystems where they are top consumers on lower trophic levels (e.g., piscivore, planktivore, benthivore fish).

The major groups of amphibians found in lakes and ponds are frogs, toads and salamanders. Some species live their whole life in freshwater whereas other species are completely terrestrial and depended on water for their reproduction. Most tadpoles have a feeding apparatus that allows them to trap bacteria, phytoplankton, and other small particles suspended in the water. Many species also graze on periphytic algae and some species even have mouth parts adapted for a predatory feeding mode. Salamanders start to feed on zooplankton but as they grow they include larger invertebrates in their diet and some species even prey on tadpoles. Tadpoles, frogs, toads and salamanders are important source of food for fish and birds.

Aquatic fish and amphibians appear to have low sensibility to glyphosate itself. Values of LC_{50} obtained in glyphosate treatments ranged from 130 mg a.i./L for the channel catfish Ictalurus punctatus to 620 mg a.i./L for the carp Cyprinus carpio. In amphibians, values of LC_{50} obtained with glyphosate (IPA salt) treatments, varied from 340 to 460 mg a.e./L in four tadpoles of Australian frogs. However, lower values where obtained in treatments with glyphosate as an acid, reporting values of LC_{50} from 82 to 121 mg a.e./L. Similar toxicity was registered for Roundup Biactive, a commercial formulation without surfactant POEA.

Wide differences were observed in the toxicity of glyphosate itself and commercial formulations. In fish, values of LC_{50} obtained with Roundup treatments ranged from 2.3 mg a.i./L for the fathead minnows Pimpehales promelas to 14.5 mg /L for Ictalurus punctatus. Treatments with Vision showed middle LC_{50} (10.42 mg a.i./L) for the rainbow trout Oncorhynchus mykiss. In addition, values of 4 and 12 mg a.i./L of Eskoba III Max (a commercial formulation with unknown surfactant) elicited the 25 % and 20% of mortality in juveniles of Pejerrey Odontesthes bonariensis and adults of Tosquero Jenynsia lineata respectively; though not lethal effects were observed in glyphosate treatments. However, much lower concentrations of Roundup have shown to cause effects in the biometry, metabolism and enzyme activities of fish. For instance, Glusczak et al. reported a significant decrease in AChE activity (enzyme presents in cholinergic synapses and motor end plates) and TBARS levels (a measure of oxidative stress) in the brain of silver catfish Rhamdia quelen exposed to 0.2 and 0.4 mg/L. Besides, significant reduction in the biometry of Piava Leporinus obtusidens was observed in treatments with 1 mg/L, eliciting a reduction of length (15%) and weight (50%). Other recent study, indicated molecular responses for the flounder Platichthys flesus treated with low doses of herbicide cocktail (< 10 μg/L of glyphosate) during a long-term contamination (62 days).

Table: Effects of glyphosate, different commercial formulations of glyphosate and POEA on fish.

VERTEBRATES	ASSESSED CHEMICAL	STUDY TYPE	ASSESSED PARAMETER	EFFECTS CONCENTRATION (mg/L) ##
Fish				
Oncorhynchus mykiss	Gly. (acid)	SST (96h)	Mortality	96h LC_{50} =**140.0**
	Roundup®			96h LC_{50} =**8.3**
	POEA			96h LC_{50} =**2.0**
Pimpehales promelas	Gly. (acid)	SST (96h)	Mortality	96h LC_{50} =**97.0**
	Roundup®			96h LC_{50} =**2.3**
	POEA			96h LC_{50} =**1.0**
Ictalurus punctatus	Gly. (acid)	SST (96h)	Mortality	96h LC_{50} =**130.0**
	Roundup®			96h LC_{50} =**13.0**
	POEA			96h LC_{50} =**13.0**
Oncorhynchus mykiss	Roundup®	SSFE (96h)	Mortality	96h LC_{50} =**52.0**
Oncorhynchus mykiss	Roundup®	SST (96h)	Mortality	96h LC_{50} =**8.5**
	POEA			96h LC_{50} =**3.2**
Oncorhynchus nerka	Roundup®	SST (96h)	Mortality	96h LC_{50} =**8.1**
	POEA			96h LC_{50} =**2.6**
Oncorhynchus mykiss	Vision®	SST (96h)	Mortality	96h LC_{50} =**10.2**
Cyprinus carpio	Gly. (acid)	SST (96h)	Mortality #	96h LC_{50} =**620.0**
Ictalurus punctatus	Roundup®	SST (96h)	Mortality	96h LC_{50} =**14.5**
Lepomis macrochirus	Roundup®	SST (96h)	Mortality	96h LC_{50} =**13.0**
Rhamdia quelen	Roundup®	SST (96h)	Metabolism & enzyme activity#	**0.2** mg/L elicited a decrease in AChE and TBARS.
Prochilodus lineatus	Roundup®	SST (96h)	Mortality & physiology #	96h LC_{50} =**13.7**
Leporinus obtusidens	Roundup®	SST (90 d.)	Biometry & enzyme activity#	**1** mg/L elicited a decrease in length and weight and AChE
Platichthys flesus	Roundup® + AMPA	SST (62 d.)	Molecular and physiology	**0.16** μg/L elicited liver injury
Odontesthes bonariensis	Gly. (IPA salt)	SST (96h.)	Mortality	Not observed lethal effect
	Eskoba III Max®			**4** mg/l elicited the 25% of mortality
Jenynsia lineata	Gly. (IPA salt)	SST (96h.)	Mortality	Not observed lethal effect
	Eskoba III Max®			**12** mg/l elicited the 20% of mortality

Comparable outcomes with treatments of Roundup and Glyphos (other commercial formulation containing POEA) were reported for amphibians. In laboratory tests, values of LC_{50} varied from 2.6 mg/L of Glyphos for tadpoles of the hylid Scinax nasicus to 11.6 mg a.e./L of Roundup for tadpoles of Litoria moorei. Middle concentrations (8 mg a.i./L) caused 100% of mortality in tadpoles of the toad Rhinella arenarum (Pérez & Miranda unpublished) exposed to Eskoba III Max, though LC values of 3.2 mg a.i./L were reported for this toad exposed to Roundup Ultra-Max. However, lower concentrations have shown significant effects in mortality and growth. For instance, Chen et al. reported 100 % of mortality in tadpoles of Rana pipiens treated with 0.75 mg a.e./L of Vision at pH of 7.5. Cauble & Wagner observed 50% of mortality for tadpoles of Rana cascadae treated with 1.94 mg a.i./L of Roundup and an earlier metamorphosis time with 1 mg a.i. /L. In addition, 2 mg a.i./L significantly reduce the survival and growth in three of five tadpoles exposed to Roundup. The same author reported in a mesocosms experiment a 100% of mortality for the Rana sylvatica and Hyla versicolor tadpoles and around 98 % of mortality for Rana pipiens and Bufo americanus tadpoles due to a direct herbicide effect. Tadpoles seem to be more sensible to commercial formulations than juveniles and adults.

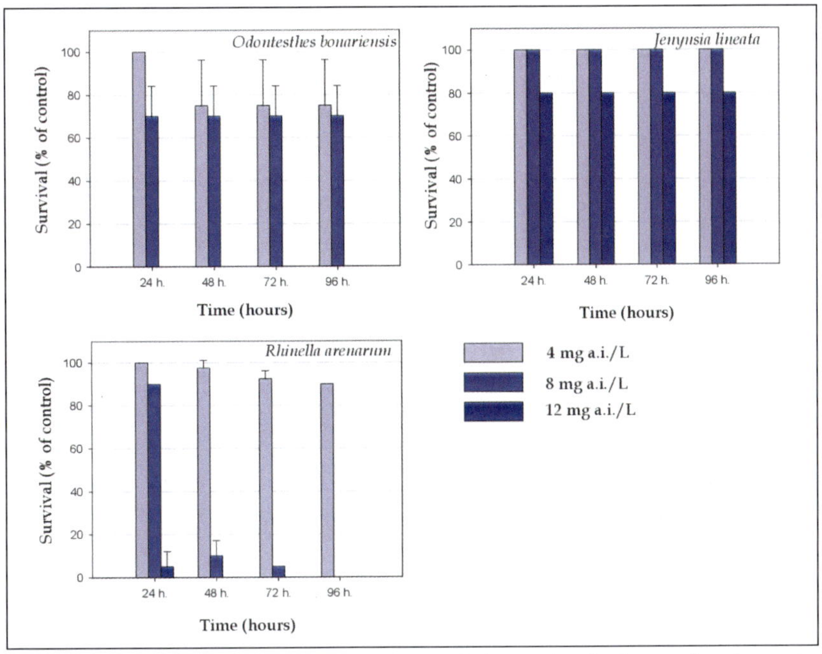

Acute lethal effects of Eskoba III Max on two fish species (Odontesthes bonariensis and Jenynsia lineata) and on tadpoles from Rhinella arenarum.

POEA itself resulted more toxic than Roundup, being this surfactant the more noxious component of several commercial formulations. Different authors concluded that the high mortality in fish and amphibian are actually due mainly to POEA surfactant and not to glyphosate itself. In the fish Pimephales promelas, the relative contribution of glyphosate acid to the toxicity of Roundup was about 30%, while glyphosate (as IPA salt) was not toxic for 5 species of Australian frogs, and therefore without any contribution to Roundup toxicity. In fish, values of LC_{50} obtained in POEA treatments varied from 1 to 13 mg a.i./L, being these values up to 4 fold more toxic than Roundup. Besides, Perkins et al., found LC_{50} values of 6.8 mg a.e./L for POEA treatments in African Xenopus laevis tadpoles, showing 1.8 fold higher toxicity than commercial formulation.

Table: Effects of glyphosate, different commercial formulations of glyphosate and POEA on frogs and toads.

VERTEBRATES	ASSESSED CHEMICAL	STUDY TYPE	ASSESSED PARAMETER	EFFECTS CONCENTRATION (mg/L) ##
-Amphibians (frogs and toads)				
Lymnodynastes dorsalis	Gly. (IPA salt)	SST (48 h.)	Morality	48h LC_{50} = > **400**
	Roundup®			48h LC_{50} = **3.0**
	Roundup® Biactive			48h LC_{50} = > **400**
	Touchdownt®			48h LC_{50} = **12.0**
Heleioporus eyrei	Gly. (IPA salt)	SST (48 h.)	Morality	48h LC_{50} => **373**
	Roundup®			48h LC_{50} = **6.3**
	Roundup® Biactive			48h LC_{50} => **427**
	Touchdownt®			48h LC_{50} = **16.1**

Crinia insignifera	Gly. (IPA salt)	SST (48 h.)	Morality	48h LC_{50} = > **466**
	Roundup®			48h LC_{50} = **3.6**
	Roundup® Biactive			48h LC_{50} = > **494**
	Touchdownt®			48h LC_{50} = **9.0**
Xenopus laevis	Rodeo®	SST (96 h.)	Morality	96h LC_{50} = **5407**
	Roundup®			96h LC_{50} = **9.4**
	POEA			96h LC_{50} = **2.7**
Scinax nasicus	Glyphos®	SST (96 h.)	Morality #	96h LC_{50} = **2.6**
Rana pipiens	Vision®	SST (8 d.)	Morality #	**0.75** mg/L elicited 100 % mortality
Rana pipiens	Roundup®	SST (16 d.)	Mortality & growth	Not observed significant effects
Rana clamitans				**2** mg/L reduce survival & growth
Rana catesbeiana				**2** mg/L reduce survival & growth
Bufo americanus				**2** mg/L reduce growth
Hyla versicolor				Not observed significant effects
Rana sylvatica	Roundup®	MES (15 d.)	Mortality & biomass #	**3.8** mg/L elicited 100% mortality
Rana pipiens				**3.8** mg/L elicited 98% mortality
Bufo americanus				**3.8** mg/L elicited 98% mortality
Hyla versicolor				**3.8** mg/L elicited 100% mortality
Pseudacris crucifer				Not observed significant effects
Rana cascadae	Roundup®	SST (42 d.)	Mortality & metamorphosis#	**1.94** mg/L elicited mortality and earlier metamorphosis times
Rhinella arenarum	Gly. (IPA salt)	SST (96h.)	Mortality	Not observed significant effects
	Eskoba III Max®			**8** mg/L elicited 100% mortality

Reviewing the available information on toxicity of glyphosate and its formulations on different groups of aquatic organisms, we have concluded that they are hazardous to the aquatic environment. Several contributions reviewed here reported significant effects of the herbicide at concentrations lower than EEC (2.6 mg a.i./L). Herbicide could be very noxious in standing waters like ponds, or in irrigation canals and impounded waters, where EEC can be reached. In these scenarios, toxicity could be exacerbated by other stressors and water characteristics (e.g. high temperature and pH, low O_2 concentration, presence of clay colloids, water hardness and other chemicals). Besides, toxicity also has showed to depend on organism life stage.

Overall, ecotoxicological sub-lethal endpoints based on behavioral traits (e.g., predator avoidance, feeding, and locomotion) and other endpoints (e.g. growth, reproduction and metabolism) seem to be more sensitive indicators of effects (i.e. reporting lower effective concentrations) and give more insights into patterns of toxicity than survivorship tests (i.e. lethality). In addition, in doses dependent effects studies, commonly results are expressed as LC_{50} or EC_{50}. However, it is not possible to predict, for instance, if the 10 % of mortality or reduction in growth (i.e at lower herbicide concentration) do not have significant effects on population and eventually in the community. On the other hand, studies focused in natural or assembled communities (e.g. microcosm and mesocosms experiments) have provided interesting and significant outcomes regarding direct and indirect herbicide effects that could not be reached in single species laboratory tests. Although these laboratory tests are an essential protocol to rapidly identify the direct impacts of pesticides on organisms, they prevent an assessment of effects on organisms embedded in their natural ecological contexts.

Glyphosate itself (as acid or salt) is generally considered to be slightly or moderately toxic to aquatic organisms (i.e., LC_{50} or EC_{50} between >1 to < 100 mg/L). However, some algae and aquatic plants showed higher sensibility, being glyphosate very toxic (EC_{50} between >0.1 to < 1 mg/L). Aquatic plants seem to be more sensitive to glyphosate than microalgae. The high toxicity of glyphosate in algae and aquatic plants is related with the mode of action of this compound (an herbicide) that interferes with plant metabolisms. On the other hand, much lower glyphosate toxicity was observed for other aquatic organisms (i.e. bacteria, protozoa, invertebrates, fish and amphibians). However, snails and worms seem to be exceptions; showing significant effects in growth, reproduction and metabolism at concentrations of < 1 mg/L of glyphosate.

Commercial formulations and specially those containing the surfactant POEA, showed higher toxicity than the active ingredient itself for all the aquatic organisms studied. Roundup showed to be up to 7 folds more noxious than glyphosate in algae and aquatic plants, up to 13 folds in protozoa, up to 42 folds in fish, up 70 folds in crustaceans, and up to 130 folds in frogs and toads. Algae and aquatic plants, showed significant effects with concentrations < 3 mg a.i./L. however, lower values were registered in studies of periphyton and micro plankton communities. Roundup concentration of 10 μg a.i./L elicited changes in marine microbial community structure and 0.13 mg a.i./L of Rodeo caused an increment in periphyton primary production. In addition, significant effects at concentrations relevant to environmental toxicity thresholds were also observed for other groups of aquatic organism. In protozoa, Euglena gracilis showed high sensibility, with 0.1 mg a.i. /L of Roundup and Avans eliciting significant sublethal effects. Different species of crustaceans showed lethal effects with values lower than 3 mg a.i./L of Roundup. In Frogs and toads, relevant concentrations of glyphosate based products (< 2 mg a.i./L) elicited lethal and sublethal effects. Fish seems to be less sensitive to commercial formulations, though some contributions showed significant sublethal effects in metabolism and enzyme activity at concentrations (< 2.5 mg a.i./L) of Roundup.

The high toxicity observed for several commercial formulation of glyphosate was generally related with the content of POEA. In protozoa and invertebrates, POEA contributed with more than the 80% of Roundup toxicity. Crustaceans showed values of EC_{50} and LC_{50} that ranged from 0.097 to 2 mg/L of POEA. In fish, glyphosate only contributed to the toxicity of Roundup with around 30%, and values of EC_{50} ranged from 1 to 13 mg/L of POEA. In frogs and toads POEA seems to be the most toxic compound in commercial formulations. Glyphosate alone (as IPA salt) was not toxic for 5 species of Australian frogs (LC_{50} > 343 mg a.e./L) and therefore contributed little to Roundup toxicity. In contrast, POEA could show lower contribution in algae and aquatic plants (as from about 46%).

Stated the hazard of glyphosate and commercial formulations of glyphosate on aquatic environments and ecological implication of the effects reviewed here, we stressed the paucity of contributions studying the effects of glyphosate on several potential endangeredaquatic species (e.g. hydroids, sponges, worms, flatworms, insects, and urodela species). We also emphasize the necessity of studies in natural communities or in assembled communities in order to evaluate direct and indirect effects upon different trophic levels.

Finally we consider that glyphosate and commercial formulations of glyphosate could have particularly significant disruptive effects to waterbodies like ponds. Ponds have been widely recognized as very important freshwater habitats. These relative small and shallow still aquatic environments are very rich in genetic and taxonomic biodiversity; they are important refuges for amphibians and also for a bewildering variety of plants and animals, including many scarce and endangered species. In addition ponds are important places for insects hatching, fish larvae and juveniles refuges and net sites for wetland birds.

Chapter 5

Integrated Weed Management

The management of weeds using multiple control tactics is known as integrated weed management. The goal of this approach towards weed control is to allow the producers to include different methods for controlling weeds in a single growing season in order to maximize the efficiency of weed control. This chapter has been carefully written to provide an easy understanding of the varied facets of integrated weed management.

Increased world population will demand more food production, which can only be achieved by increasing crop yields and applying a sustainable approach, i.e. more production with rational use of available resources, which also implies responsible use of land and water and enhanced food diversity. Efforts are needed to reduce crop losses due to pests through the implementation of Integrated Pest Management (IPM) (resistant crop varieties, rational use of pesticides, biocontrol and better cultural practices) without harmful side-effects.

Among the pests, weeds are considered an important biotic constraint to food production. Their competition with crops reduces agricultural output (quantity and quality), and increases external costs by spreading them across farm boundaries. It is also a major constraint to increased farmers' productivity, particularly in developing countries. The labour needed for hand- weeding one hectare of land is not within reach of the small farmer's family.

An integrated weed management approach to land management combines the use of complementary weed control methods such as grazing, herbicide application, land fallowing, and biological control. The resulting combinations provide the best possible solutions to weed problems for land managers. By studying the impact of each of the above methods individually as well as in combination, sustainable management systems can be devised to suit different regions and catchment areas.

Herbicide Application

Herbicides applied at the wrong time of the year can be ineffective. This wastes both time and money for the farmer. Mature plants have already produced large amounts of seeds which simply add to the soil seed bank reserves.

Land Fallowing

The strategic fallowing of land may provide windows of opportunity for perennial pastures to establish. This helps them out-compete weed species. A common characteristic of weedy species is that they are generally first level colonisers. That is, when other vegetation is removed by overgrazing, clearing or ploughing and the ground is left bare, weeds establish quickly and reproduce effectively to maintain that niche. Trials where land is fenced off show that some weed species have difficulty competing with other vegetation.

Biological Control

Biological control agents are generally most effective when established in gullies and rocky knolls of hills. These areas are usually inaccessible and too costly to spray so they can provide a safe haven for agents to retreat to from pastures when there is too much disturbance, such as grazing.

Herbicide Tolerant Crops

The discovery of herbicide-resistant weeds in the early 1970s triggered an interest in mimicking this unintentional development for use in crop breeding. The concomitant progress in molecular genetics made it possible to incorporate resistance genes from unrelated organisms into an otherwise susceptible crop. In other words, we were now able to adapt the biology of the crop to the chemistry of an herbicide, whereas we previously had to adapt chemistry to biology. It must, however, be noted that herbicide-resistant crops (HRCs) were first produced by methods of traditional breeding, whereas the major current HRCs have been produced by genetic engineering, the technology which has unintentionally placed these crops in a fierce debate between those in favour, and those against, the introduction and commercial use of genetically modified (GM) crops.

Components of an IWM Plan

The goal of IWM is to incorporate different methods of weed management into a combined effort to control weeds. Just as using the same herbicide again and again can lead to resistance, reliance on any one of the methods below over time can reduce its efficacy against weeds. Two major factors to consider when developing an IWM plan are (1) target weed species and (2) time, resources, and capabilities necessary to implement these tactics.

While it is wise to be a good steward of herbicide technology, through the use of PRE and POST herbicide applications or tank mixes, IWM requires the use of tactics beyond herbicides. For example, using these herbicide application practices along with a winter cover crop or harvest weed seed control (HWSC) and prevention methods would be considered IWM.

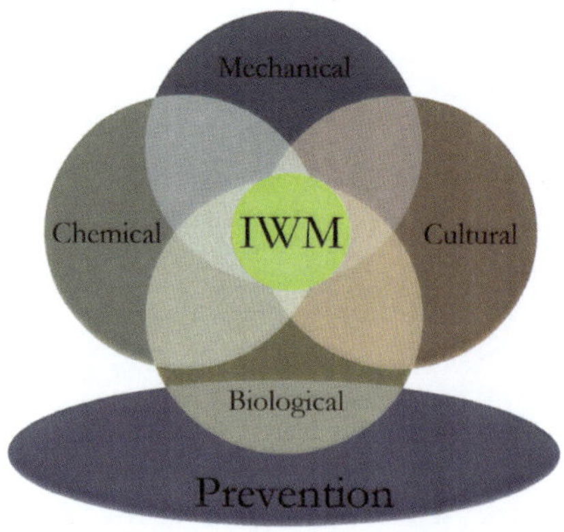

IWM is composed of mechanical, cultural, chemical and biological tactics.

Categories of IWM Practices

1. Prevention: Prevention is one of the first steps of weed management. This category is unlike the others in that it focuses on keeping weeds out of the field or spreading within a field.

Growers can incorporate this tactic by:

- Avoiding crop seed, manure, and other inputs that are contaminated with weed seeds.
- Cleaning equipment, including combines (combine cleaning methodology), that could transport weed seeds between fields.
- Preventing weeds from producing seeds in the field but also in ditches, fencerows, and other nearby non-crop areas.
- Scouting for weeds in a timely manner.
- Proceeding with caution when purchasing used farm equipment or using rented land.

Horseweed seed on bush hog after mowing a weedy field. These seeds can easily spread to otherfields.

Feeding combine with straw bales for deep cleaning.

2. Cultural: A healthy, vigorous crop is the best weed control. Cultural practices are designed to give the crop a competitive advantage over weeds.

Growers can incorporate this tactic by:

- Reduced row spacing so the crop can reach canopy more quickly to shade out weeds.
- Crop rotation to prevent weeds from adapting to the weed control tactics common in any one crop.
- Nutrient management to allow optimum crop uptake while denying weeds access to nutrients.
- Cover crops to compete with weeds for space, sunlight, nutrients, and water.

- Altered planting dates to give the crop a head start or allow for a flush of weed germination that can be controlled before planting.
- Crop variety selection to ensure crops have the utmost competitive advantage against weeds.

Horseweed suppression from a cover crop mixture compared to an area where no cover crops were planted.

3. Chemical: Herbicides are an integral part of most weed management plans and will continue to be so, even in IWM programs.

Good management practices for applying herbicides include:

- Timely scouting.
- Proper weed identification and awareness of what herbicide-resistant weeds are in the area.
- Correct herbicide application, meaning applying the appropriate product at the right rate and at the right time.
- Maximized diversity through the use of tank mixes herbicides with different, effective sites of action (SOA) and by rotating herbicides throughout the season whenever possible.
- Plan ahead across seasons to avoid using herbicides with the same SOA repeatedly.

Application Timing	Herbicide	SOA number	Total SOA	Effective SOA
PRE	Bicep	5 + 15		
POST	Atrazine + glyphosate	5 + 9	2	2

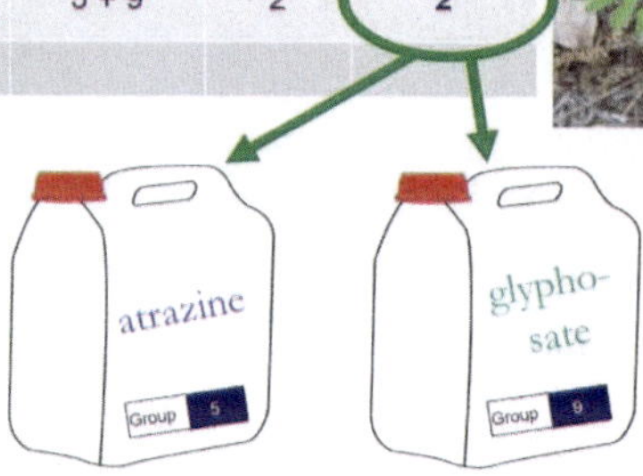

Understanding the concept of herbicide Site of Action (SOA) is key to effectively managing herbicide resistance.

4. Mechanical: Mechanical weed management focuses on physical practices that disrupt germination and destroy plant tissue.

Growers can incorporate this tactic by:

- Hand-pulling
- Tillage
- Burning
- Mowing
- Robotic weeding machines
- Harvest weed seed control, which reduces the input of weed seeds into the soil seedbank by destroying or removing seeds retained on the weeds at the time of harvest.

Hand pulling escaped weeds is critical to prevent seeds from entering the soil seed bank, particularly for herbicide resistant weeds such as Palmer amaranth.

Windrow burning, a form of harvest weed seed control, is an excellent tactic to prevent weed seeds from entering the soil seed bank.

Harrington Seed Destructor: Two mills destroy weed seeds contained in the chaff portion that comes out from the combine.

5. Biological: This tactic uses living organisms to target weeds including bacteria, fungi, or insects that have a preference for a certain weed species. This tactic is arguably the least used of all tactics but is the subject of much research. Cover crops can be considered a biological control tactic.

Ecological Importance of IWM

In most ecosystems, herbicides have become one of the most important components in weed control. There are two reasons to explain the increased use of herbicides, the first being the widespread adoption of high-yielding varieties which created economic incentives for farmers to reduce weed infestation; and the second is the availability of cheap herbicides, indicating that the cost of weed control by herbicides in wet-seeded rice is less than one-fifth of the cost of a single hand-weeding in Illoilo, Philippines. Similar situations exist in West Java, Indonesia and the Mekong Delta, Vietnam.

Because of the availability of cheap herbicides, it is expected that herbicide usage will continue to increase, both in developed countries, and even in developing countries, where herbicides are currently used sparingly and farm wages are relatively low. However, this does not indicate a lack of importance for hand-weeding. Manual weeding is still the dominant weed control method in many parts of Asia, since management options for weed control are limited under diverse agro-ecological conditions.

All the recommended herbicides commonly used are known to be very safe not only for humans and cattle, but also for the environment when they are used properly. Sulfonylurea type herbicides have been intensively used in far-east Asia since 1990 because of their high efficacy against a broad spectrum of paddy weeds even at extremely low doses. However, intensive and repeated application of this type of herbicide has resulted in several negative effects, as follows:

- Evolving resistant weeds.
- Residual effects on the following crops.
- Disappearance of some susceptible weeds such as Brasenia schreberi and Sagittaria aginashi, which affects weed biodiversity.

All these factors may well provide sufficient reason to attract public concern and anxiety regarding the negative effects of herbicides that might originate from intensive herbicide application in the environment. In this regard, an alternative to such a heavy dependence on herbicide is needed. Such an alternative might be found in the use of integrated weed management, which can reduce herbicide use in different cropping systems.

Organic Agriculture

During recent years, organic agriculture has hugely increased in many regions of the world, and this trend has not yet ceased. The total area of organically grown crops is estimated to be around over 8 million ha. Organic agriculture should be seen as a process depending highly on naturally-occurring biological processes. Therefore, the objective is to stimulate these processes to obtain maximum suppression of pest problems. Obviously, the use of chemical pesticides is contrary to the concept and the practice of organic agriculture, and human intervention in the process should therefore be carried out only very carefully.

Weed management in organic agriculture is not an easy task, particularly in areas where labour for hand-weeding is short or not affordable. However, the principle should be the same as in any conventional cropping system, i.e. weed competition needs to be prevented in order to obtain maximum crop yields. This necessarily implies weeding with non-chemical materials but, this has to be carried out exactly at the right time to eliminate weeds during the so-called 'critical period of weed competition. Organic systems also require the use of preventive methods before growing the crop and to establish a reasonable crop rotation. Stale seed bed preparation in order to kill the weeds mechanically or manually is a very good option to delay the start of weed competition. The use of cover crops and green manure, in addition to increasing soil fertility, may help to control some weed species. The most common methods used to prevent weed competition in organically grown crops are high seeding rates, narrow seed spacing/cross seeding, and companion cropping with small-seeded legumes. Rotational sequences for crops grown in tropical and sub-tropical areas should be developed to guarantee vigorous growth of the crops as well as fewer pest problems, including weeds.

Weed Control in Conservation Agriculture

The so-called Conservation Agriculture (CA) is gaining a lot of recognition among farmers all over the world. Grain crops as wheat, barley, maize, rice and soybean are grown in some parts of the world under this system, which consists of the judicious use of crop rotation with minimum or zero tillage, including the use of green manures and cover crops. This approach is beneficial to effectively protect and increase soil fertility. It is wrong to identify conservation agriculture as the practice of zero or minimum tillage. In fact these procedures are part of the system, but if they are implemented in areas of monocropping then this cannot be considered conservation agriculture. The use of herbicide-resistant crops (HRCs) combined with the application of broad-spectrum herbicides make the process of conservation agriculture easier, but it also runs the risk of bringing about new weed problems, either by a shift in the weed populations or the presence of species able to evolve resistance to the herbicides in use. Again it is not always necessary to use HRCs and herbicides, particularly when the system has been practised for ten years or more.

Although this development is real, several weeds still remain which seriously affect the production of a number of crops such as rice, with its incidence of red/weedy rice, and Echinochloa spp., parasitic Orobanche weeds found in sunflower, faba beans and solanaceous crops; parasitic Striga spp. in cereals in Africa south of the Sahara, and Imperata cylindrica in many areas in Africa.

The switch to zero tillage or direct seeding practices concerns many producers when it comes to controlling weeds. The loss of tillage as a method of weed control means that producers must adjust crop rotations, herbicide use, and other cultural practices to compensate. Perennial weeds may become a serious problem to overcome, and there is a need to implement additional cultural methods, such as the use of covers. Some annual weeds, such as wild oat and volunteer canola, also grow well under zero-till conditions.

Weeds exist in many different forms and with different life spans; there are annual, biennial, and perennial weeds. Weeds are not always bad, and in low density will result in no, or small yield losses. Heavy manifestations of weeds, or establishment of perennial weeds, can result in large yield losses or can even take land out of production until the weeds are controlled. Weeds can also reduce yield quality and can be toxic when ingested by animals or humans. Weeds can also cause environmental damage and loss of agricultural biodiversity, by competing for inputs. Integrated Weed Management aims at:

- Preventing weeds from spreading by:
  - Cleaning farm machinery and vehicles before transporting, to avoid risk of spreading weeds;
  - Cleaning the hair and feet of animals before moving to new areas;
  - Controlling the weeds in feed and bedding grounds;
  - Using only well stored and rotted manure (4-5 months), possibly improve decomposition;
  - Making sure that soil disturbances are immediately reseeded
  - When possible, practicing weed control on all aspects of the farm, including irrigation canals, drainage ditches, fence lines, stockyards, and farm roads.
- Improving knowledge of the identification and effects of different types of weeds.
- Monitoring and map the spread of weed populations and the resulting damage
- Making control decisions based on full knowledge of potential damage, cost of control methods, and the environmental impact of the control strategy.
- Using combinations of (preferably biological) weed control strategies to reduce the weed populations, these can include: winter cover crops, mulching, crop rotations, natural competition (e.g. ryegrass), livestock grazing, proper seedbed preparation, selecting locally adapted varieties, proper fertilizer application, stimulating bio-control by insects, mowing, hand weeding, and try to avoid, when possible herbicide application, tillage, and burning.
- Evaluate and monitor the effectiveness and (environmental) effects of control strategies.

IWM in Some Important Crops

Rice

Critical period of weed control	20-30 DAT
Cultural method	1. Hand weeding 2. Hand pulling 3. Pudding 4. Flooding
Mechanical method	1. Weeder (Float) 2. Conoweeder/Rotary weeder
Chemical method	Apply pendimethalin 1.0kg/ha on 5 days after sowing or Pretilachlor + safener (Sofit) 0.45kg/ha on the day of receipt of soaking rain followed by one hand weeding on 30 to 35 days after sowing.
Biological method	1. Hirsch – Manniella spinicaudata is a rice root nematode which controls most upland rice weeds. 2. Azolla
Remarks	I. Substitution and preventive method: a. Stale seed bed technology b. Land preparation c. Water management

Wheat

Critical period of weed control	15 – 30 DAS
Cultural method	1. Hand Hoeing 2. Inter cultivation 3. Criss-cross sowing
Chemical method	1. 2, 4D (1 – 1.5 kg ai/ha) 2. Mixture of Isoproturan (0.75 kg ai/ha) and 2, 4D (0.4 kg ai/ha) during 30- 35 DAS
Remarks	II. Complimentary weed control methods 1. Cultivars 2. Seedling age /planting method 3. Fertilizer management 4. Cropping system

Sorghum

Critical period of weed control	21 – 42 DAS
Cultural method	1. Hoe and hand weed on the 10th day of transplanting if herbicides are not used. 2. Hoe and weed between 30 - 35 days after transplanting and between 35 - 40 days for a direct sown crop.
Chemical method	1. Apply the pre-emergence herbicide Atrazine 50 WP - 500 g/ha on 3 days after sowing as spray on the soil surface 2. If pulse crop is to be raised as an inter-crop in sorghum do not use Atrazine.

Maize

Critical period of weed control	Maize 2 to 6 Weeks.
Cultural method	One hand weeding on 40-45 days after sowing.
Chemical method	1.Pre-emergence application of Atrazine (1-2 kg ai/ha) 2. Combined application of Alachlor (2 kg/ha) and atrazine (1kg/ha) is more effective and have wider spectrum of ontrol.

Groundnut

Critical period of weed control	Upto 45 days
Cultural method	After 35 - 40 days one hand weeding may be given.
Chemical method	Alachlor (1-5 kg/ha) – pre emergence application.

Sunflower

Critical period of weed control	4-6 weeks
Cultural method	Hoe and hand weed on the 15th and 30th day of sowing and remove the weeds. Allow the weeds to dry for 2 - 3 days in the case of irrigated and then give irrigation.
Chemical method	Apply Fluchloralin at 2.0 l/ha before sowing and incorporate or apply as pre-emergence spray on 5 day after sowing followed by irrigation or apply Pendimethalin as pre-emergence spray 3 days after sowing.

Cotton

Critical period of weed control	First 45 days.
Cultural method	1. One hand weeding on 45 DAS will keep weed free environment upto 60 DAS. 2. Hoe and hand weed between 18th to 20th day of sowing, if herbicide is not applied at the time of sowing.
Chemical method	Diuron(0.5–1.5kg/ha),Monuron(1-1.5kg/ha), Fluchloralin (1-1.5 kg/ha) applied as preemergence/preplanting.

Pulses

Critical period of weed control	First 30-35 days.
Cultural method	1. One hand weeding on 30 days after sowing gives weed free environment throughout the crop period. 2. If herbicides are not applied give two hand weedings on 15 and 30 days after sowing.
Chemical method	Fluchloralin (1-1.5 kg/ha), Pendimethalin (0.5-1.0 kg/ha) as Pre emergence (preplanting incorporation).

Tobacco

Critical period of weed control	First 9 weeks.
Chemical method	Fluchloralin (2-3 kg/ha), Pendemethalin (1-1.5 kg/ha) as pre-emergence application.

Sugarcane

Critical period of weed control	4 to 5 months.
Cultural method	Remove the weeds along the furrows with hand hoe.
Mechanical method	If herbicide is not applied work the junior-hoe along the ridges on 25, 55 and 85 days after planting for removal of weeds and proper stirring.
Chemical method	1. Pre-emergence herbicides like atrazine (2 to 3 kg/ha) Simazine (2 to 3 kg/ha), Alachlor (1.3 to 2.5 kg/ha) etc., will generally last for 8 to 12 weeks. 2. To obtain best results sequential application of Preemergence and post emergence herbicides or post emergence herbicides like Glyphosate (0.8 to 1.6 kg/ha) Paraquat (0.4 to 0.8 kg/ha).

Integrated Weed Management Plan

It would be useful to prescribe one specific integrated weed management plan to fit each crop in each growing area, but this is not realistic. Every farm, indeed every field, needs to be evaluated on an individual basis to determine the most efficient and economical weed control programme.

Farmers need to look ahead and plan their weed control programmes for as long as practical, possibly 5-10 years, but the longer the better. Plans can always be modified as new technologies and practices become available. The aim is to craft an appropriate weed management programme that will not rely solely on one herbicide mode of action over successive years. In this way the potential risk for the selection of herbicide resistant weeds and weed shifts can be minimised. To develop an integrated weed management programme it is necessary to think strategically about how best to utilise all available weed control methods in combination to give the best overall result for each crop and each rotation. A checklist of options is a useful tool for farmers to develop.

This topic points out some of the factors that must be considered when developing an integrated weed management plan for conventional or herbicide tolerant crops. This ordered approach has been valuable to others in developing integrated weed management plans.

Purpose

The farmer needs to define the purpose of the weed management plan and what it will achieve. This requires creating long and short-term goals for the farm's integrated weed management programme. These goals will be established based on the level of weeds in the fields and the importance of these to local crop production.

Weed Management Areas

The weed management areas need to be defined in terms of the specific boundaries of each area to be managed and the background to these areas. A management area may be anything from a portion of one field to a group of fields that have similar weeds and will share a similar resistance management strategy. The management area information should include a list of the soil types in the areas; the general topography; vegetation cover; potential impact from adjacent vegetation; etc., for each area.

Problem Weeds

The farmer needs to identify the weed problems for each of the management areas. This should include listing, for each field, the three to five key target weeds and sketching maps to show the location of known weed infestations in each field. Help with the identification of weeds can be obtained on the web, or from local advisors. Importantly, the types and amounts of weedy species do not stay static. Instead, they fluctuate based on environmental factors and previous management practices. As such, farmers need to be aware of the changes and keep up-to-date with the current weed problems in production fields.

Effectiveness of Control Measures

Before planning the next phase of control, the farmer needs to review, for each management area, the effectiveness of previous weed control efforts and to list any known or suspected herbicide resistant weeds. It is useful to record the cropping and herbicide application history for each field in the management area and to note the mode of action of each herbicide that has been used.

Planned Crop Rotations

Looking forward, the farmer should list the potential crop rotations planned for each field and note any potential volunteer issues that might arise for each crop rotation. Five to 10 year planning is valuable. These plans can be changed and updated when needed.

Control Strategies and Resources

It is useful to compile a list of all practical weed management strategies and methods that are available for each field, including appropriate weed control methods in the categories of prevention, cultural, mechanical, chemical, and biological weed control practices. This list should include the known effectiveness of each practice against the target weeds in each field and any restrictions, limitations, or drawbacks that may apply to each weed control strategy (e.g., herbicide carryover).

This planning documentation should include the equipment and resources needed, and the cost, for each weed control practice.

Weed Management Plans

Using the information gathered above the farmer needs to prepare a weed management plan for each management area by evaluating the most appropriate weed control practices for each weedy area. The final choice of weed management plan for each area needs to take into consideration the effectiveness and cost of the practices, the timing and the feasibility of implementing each plan with resources that are available to the farmer. In addition, the management plans should include mechanisms for the prevention of new infestations and the avoidance of the development of resistance in weeds. The management of volunteer crop plants will need to be considered and safety requirements (human and crop) for the various components of the management plan, especially the consideration of practices to minimise long term, negative environmental impact.

Implementation

With a well informed, well researched and well reviewed plan the farmer is ready to implement weed management for his fields. Once the plans are initiated, it will be important to continue to scout the fields to assess the effectiveness of the weed control practices. Accurate records are essential to be able to evaluate effectiveness. It is best to review the effectiveness on an ongoing basis so that changes can be implemented to address what is happening in the fields. If a review of the data indicates that the weed management plan needs to be modified, these changes should be planned and implemented as quickly as possible.

Finally, prior to implementing the weed management plans, the farmer would benefit from a critical review of what is planned and any revisions, if these are deemed necessary. The critical review can be guided by the following questions:

- Do the management plans incorporate sufficient different modes of action to avoid the development of herbicide resistant weeds or weed shifts?
- Will the management plans ensure that volunteer crop plants can be controlled effectively in the subsequent crop rotations that are detailed in the plan?
- Do the management plans involve the use of weed control methods other than herbicides?
- Have all the available cultural control practices been evaluated and have those that are feasible been included in the management plans?

Importance of Keeping Accurate Records

Record keeping is important and is an essential practice for the proper management of herbicide tolerant crops. Without accurate records costly mistakes in application of herbicides may be made and it may be difficult to determine if herbicide tolerant volunteers will be an issue in subsequent crops. Records should detail the whole production process from seed purchase to the final sale and delivery of the harvested crop. Some of the most important records to keep are:

- Lists of problem weeds including their prevalence in key areas.
- Effectiveness of control measures.
- Past and planned crop rotations.
- Control strategies with resources needed and resources available.
- Weed management implementation plans.
- Monitoring activities and problems identified.

- Corrective actions.
- Reviews of IWM and revisions.

Not all farmers have the time and resources for detailed record keeping, but growers are encouraged to devise documentation systems that are practical for their needs. Examples of recording forms for integrated weed management are included in this manual as a reference for growers wishing to develop their own documentation systems.

References

- Integrated-weed-management, managing-ecosystems, crops, agriculture: fao.org, Retrieved 8 April, 2019
- What-is-integrated-weed-management: integratedweedmanagement.org, Retrieved 15 August, 2019
- Agri-weedmgt-iwmintegratedweedmgt: tnau.ac.in, Retrieved 12 July, 2019
- Implementing-Integrated-Weed-Management-for-Herbicide-Tolerant-Crops: croplife.org, Retrieved 20 January, 2019

Permissions

All chapters in this book are published with permission under the Creative Commons Attribution Share Alike License or equivalent. Every chapter published in this book has been scrutinized by our experts. Their significance has been extensively debated. The topics covered herein carry significant information for a comprehensive understanding. They may even be implemented as practical applications or may be referred to as a beginning point for further studies.

We would like to thank the editorial team for lending their expertise to make the book truly unique. They have played a crucial role in the development of this book. Without their invaluable contributions this book wouldn't have been possible. They have made vital efforts to compile up to date information on the varied aspects of this subject to make this book a valuable addition to the collection of many professionals and students.

This book was conceptualized with the vision of imparting up-to-date and integrated information in this field. To ensure the same, a matchless editorial board was set up. Every individual on the board went through rigorous rounds of assessment to prove their worth. After which they invested a large part of their time researching and compiling the most relevant data for our readers.

The editorial board has been involved in producing this book since its inception. They have spent rigorous hours researching and exploring the diverse topics which have resulted in the successful publishing of this book. They have passed on their knowledge of decades through this book. To expedite this challenging task, the publisher supported the team at every step. A small team of assistant editors was also appointed to further simplify the editing procedure and attain best results for the readers.

Apart from the editorial board, the designing team has also invested a significant amount of their time in understanding the subject and creating the most relevant covers. They scrutinized every image to scout for the most suitable representation of the subject and create an appropriate cover for the book.

The publishing team has been an ardent support to the editorial, designing and production team. Their endless efforts to recruit the best for this project, has resulted in the accomplishment of this book. They are a veteran in the field of academics and their pool of knowledge is as vast as their experience in printing. Their expertise and guidance has proved useful at every step. Their uncompromising quality standards have made this book an exceptional effort. Their encouragement from time to time has been an inspiration for everyone.

The publisher and the editorial board hope that this book will prove to be a valuable piece of knowledge for students, practitioners and scholars across the globe.

Index